VIROLOGY RESEARCH PROGRESS

AEDES AEGYPTI

ECOLOGY, CONTROL AND TRANSMISSION OF DISEASE

VIROLOGY RESEARCH PROGRESS

Additional books and e-books in this series can be found on Nova's website under the Series tab.

VIROLOGY RESEARCH PROGRESS

AEDES AEGYPTI

ECOLOGY, CONTROL AND TRANSMISSION OF DISEASE

VALÈRIE MEGENS
EDITOR

Library of Congress Cataloging-in-Publication Data

ISBN: 978-1-53618-197-5

Published by Nova Science Publishers, Inc. † New York

CONTENTS

Preface		**vii**
Chapter 1	*Aedes aegypti*: Ecology, Control and Transmission of Disease *Stalin Vilcarromero, Thomas Iwanejko, Carlos Tong-Rios, Freddy Alava, Scott R. Campbell and Raul Chuquiyauri*	**1**
Chapter 2	Review on *Aedes aegypti:* Arboviral Disease Transmission and Control *Edith Nonye Nwankwo*	**33**
Chapter 3	Entomopathogenic Fungi: Applications to the Biological Control of Culicidae Vectors of Disease *Márcia Jordana F. Macêdo, Nara Juliana S. Araújo, Luis Pereira-de-Morais, Yedda Maria Lobo S. Matos, Henrique Douglas M. Coutinho, Francisco Assis B. Cunha and Maria Flaviana B. Morais-Braga*	**77**
Index		**121**

PREFACE

The *Aedes aegypti* mosquito is a recognized vector for dengue, chikungunya, and Zika arboviruses, and has had a significant dispersion in recent years across the southern hemisphere.

It is a known nuisance species in the United States and is believed to have been brought to the new world on ships used for European exploration and colonization.

In recent years, the use of entomopathogenic fungi has proven a promising tool for the biocontrol of Culicides that threaten public health. As such, the authors systematically review studies that evaluated the main entomopathogenic genera used in the biological control of these vectors.

Chapter 1 - The *Aedes aegypti* mosquito is a recognized vector for dengue, chikungunya, and Zika arboviruses, and has had a significant dispersion in recent years across the southern hemisphere. Still, even now, the *Aedes aegypti* is reported to be expanding into regions with temperate climates such as European countries and parts of North America. The impact on public health has been tremendous in terms of morbidity and mortality. Environments that favor more adult mosquitoes and transmission of disease (rapid urban growth, lack of adequate water supply, and lack of solid waste disposal) are associated with high rates of mosquito bites and risk of infection. In the peridomestic environment, stored water and discarded non-biodegradable containers accumulate rainwater and can become a habitat for the development of the immature *Aedes aegypti*. It is an efficient vector for arboviruses chiefly due to its biting patterns. The vector control measures against the growing immature *Aedes aegypti* have not been successful. The indices for immature mosquitoes are not demonstrated to be a reliable predictor for human dengue infection risk. Indoor space spraying, a method to control for the adult *Aedes aegypti* is effective during outbreaks; however, when using such chemical options, resistance to insecticide is a threat and a growing problem. Recently, the interventions of release insects with dominant lethality (RIDL) and Wolbachia have been developed to fight against dengue fever and Zika virus. These strategies have proven to reduce the incidence in many clinical trials, and more studies are being conducted to determine the long-term impact and sustainability.

Chapter 2 - *Aedes aegypti* Linnaeus (Diptera: Culicidae) is a tropical mosquito and one of the most efficient mosquito vectors for the native African arboviruses. It is believed to have spread to other continents through trade and other human aid dispersal and now is distributed worldwide. It is a known nuisance species in the United States for centuries and is believed to have been brought to the new world on ships used for European exploration and colonization. Traditionally, *Ae. aegypti* was found in forested areas using tree holes as habitats. As a domestic breeder, it was initially found in breeding places on sailing ships from where it was distributed to all parts of the world. Adult *Aedes* mosquitoes show specific selection for sources of water or container types in which they lay their eggs. Hence, the species has a unique and specific environmental requirement for maintenance of its life cycle. The mosquito bites primarily during the day and is most active for almost two hours after sunrise and several hours before sunset, but it can bite at night in well lit areas. *Aedes* mosquito is known to bite people without being noticed because it approaches from behind and bites on the ankles and elbows, thus is referred to as an aggressive biter. It prefers biting people but it also bites dogs and other domestic animals, mostly mammals. It therefore represents a major threat to public health and inflicts terrible and unacceptable public health burden to humankind. This is due to its ability to transmit numerous human and animal pathogens such as Yellow fever, Dengue, Chikungunya and Zika viruses. *Aedes* mosquito control operations range from the use chemical control (larviciding or adulticiding), botanical control, bio-control, source reduction, mosquito traps among others. These control strategies are accomplished using pesticides, plant extracts and oil, biological-control agents, habitat modification and trapping. These are targeted to reduce nuisance mosquito cause to people around homes or in parks, recreational areas and their impact on real estate values, tourism, livestock and public health.

Chapter 3 - In recent years, the use of entomopathogenic fungi has been a promising tool for the biocontrol of Culicides that threaten public health. The authors systematically reviewed (2007-2018), studies that evaluated the main entomopathogenic genera used in the biological control of these vectors. Among the genera reported in this review, *Metarhizium* and *Beauveria* stand out as the best investigated over the years, being considered highly virulent, mainly, against species of the genera *Aedes* and *Anopheles*, in their larval and adult stages. While *Culex* spp. remains little researched. From the remaining entomopathogens, the oomycete *Leptolegnia chapmanii*, appears as a promising candidate in the infectivity and death of *Ae. aegypti*.

In: *Aedes aegypti*
Editor: Valèrie Megens

ISBN: 978-1-53618-197-5

Chapter 1

AEDES AEGYPTI: ECOLOGY, CONTROL AND TRANSMISSION OF DISEASE

Stalin Vilcarromero*[1,*]*, MD, Thomas Iwanejko*[2]*,
Carlos Tong-Rios*[3]*, Freddy Alava*[3]*,
Scott R. Campbell*[4]*, PhD
and Raul Chuquiyauri*[5]*, MD, PhD
[1]Stony Brook University, New York, NY, US
[2]Suffolk County Vector Control, NY, US
[3]Loreto Regional Directorate of Health, Vector Control, Iquitos, Peru
[4]Suffolk County Department of Health Services, NY, US
[5]Medical Care Development International, MD, US

ABSTRACT

The *Aedes aegypti* mosquito is a recognized vector for dengue, chikungunya, and Zika arboviruses, and has had a significant dispersion in recent years across the southern hemisphere. Still, even now, the *Aedes aegypti* is reported to be expanding into regions with temperate climates

* Corresponding Author's E-mail: stalinvil@gmail.com.

such as European countries and parts of North America. The impact on public health has been tremendous in terms of morbidity and mortality. Environments that favor more adult mosquitoes and transmission of disease (rapid urban growth, lack of adequate water supply, and lack of solid waste disposal) are associated with high rates of mosquito bites and risk of infection. In the peridomestic environment, stored water and discarded non-biodegradable containers accumulate rainwater and can become a habitat for the development of the immature *Aedes aegypti*. It is an efficient vector for arboviruses chiefly due to its biting patterns. The vector control measures against the growing immature *Aedes aegypti* have not been successful. The indices for immature mosquitoes are not demonstrated to be a reliable predictor for human dengue infection risk. Indoor space spraying, a method to control for the adult *Aedes aegypti* is effective during outbreaks; however, when using such chemical options, resistance to insecticide is a threat and a growing problem. Recently, the interventions of release insects with dominant lethality (RIDL) and Wolbachia have been developed to fight against dengue fever and Zika virus. These strategies have proven to reduce the incidence in many clinical trials, and more studies are being conducted to determine the long-term impact and sustainability.

INTRODUCTION

The *Aedes aegypti* mosquito, a recognized dengue, chikungunya, and Zika virus vector, has had a significant dispersion in the southern hemisphere. Still, even now, the *Aedes aegypti* is reported in regions with temperate climates such as European countries and parts of North America. The impact on public health has been tremendous in terms of morbidity and mortality. Knowledge of the ecology and transmission and control of this mosquito is very important. It will allow us to know the most important factors that promote the development of its life stages, the factors involved in disease transmission, so that we can define the vector control measure. Although some vaccines are now available for mosquito-borne diseases, some have not shown the expected efficacy, due to its dependence on immune factors and the predominant serotypes. Chemical control through the use of insecticides has been widely used and has served to lessen the impact of outbreaks or epidemics. And It is of great relevance to know and use them appropriately. Insecticide resistance to pesticides has emerged, and it is a very serious problem. New genetic and biological control techniques have been developed, and we must better understand their use in considering a balance between their benefits and the uncertainties that still exist for those strategies. The objective of this chapter is to review these concepts in a concrete and simple way.

1. Ecology

1.1. Distribution

Aedes aegypti (Diptera: Culicidae) is a mosquito vector of medical importance. This specie has a cosmotropical distribution between the 37° South and 35° North latitudes. *Aedes aegypti* is found throughout most tropical to subtropical regions of the World, where it exhibits a preference for human habitats within close proximity to standing water. It is known that the ancestor of *Aedes aegypti*, *Aedes formosus* was native to Africa and lived in the wild forest, inhabiting tree hollows in juvenile stages and feeding on blood from non-human animals in the adult stage (Powell and Tabachnick 2013). *Aedes aegypti* is believed to have migrated to the New World during the fifteenth to seventeenth centuries aboard ships in the process of colonization and slave trade related transportation. The timing of *Aedes aegypti* colonization of Asia is likely during late nineteenth century when dengue fever was first reported; and, importantly, in urban settings due to the arrival of the only urban dengue vector, *Aedes aegypti* (Smith 1956) (Powell and Tabachnick 2013). After this dispersion to new areas, it's disappearances or marked reductions in some regions such as Mediterranean region, Black Sea, and Atlantic Ocean or the America region would have been attributed to establishment of appropriate water infrastructure and improved hygiene through development of piped water systems (Holstein 1967) and the use of insecticides such as dichlorodiphenyltrichloroethane (DDT), as part of malaria control (1970). The spread of *Aedes aegypti* to Australia, Europe, and Southeast Asia occurred in the last 30 years (Lwande, Obanda et al. 2015) is an indicator of the effect of climate change, and it's adaptations to survive lower temperatures and higher altitudes. Other factors that have contributed substantially to the tremendous development and spread of *Aedes aegypti* are: a rapid urban growth with a high human densities, lack of adequate water supply, lack of solid waste disposal and substandard housing, human movement (Stoddard, Morrison et al. 2009), and transportation. It has been shown how *Aedes aegypti* is able to develop all their stages of life not only in riverboats and terrestrial transportations, but also to spread around the villages located along the route (Guagliardo, Morrison et al. 2015).

1.2. Life Cycle and Ecology

In tropical and subtropical regions, environmental factors (e.g., rainfall, temperature, and relative humidity) favor its life cycle (Marinho, Beserra et al. 2016) (Eisen, Monaghan et al. 2014). Life stages of *Aedes aegypti* mosquitoes are eggs, larvae, pupa, and adults. (See Figure 1). Some characteristics of these stages are:

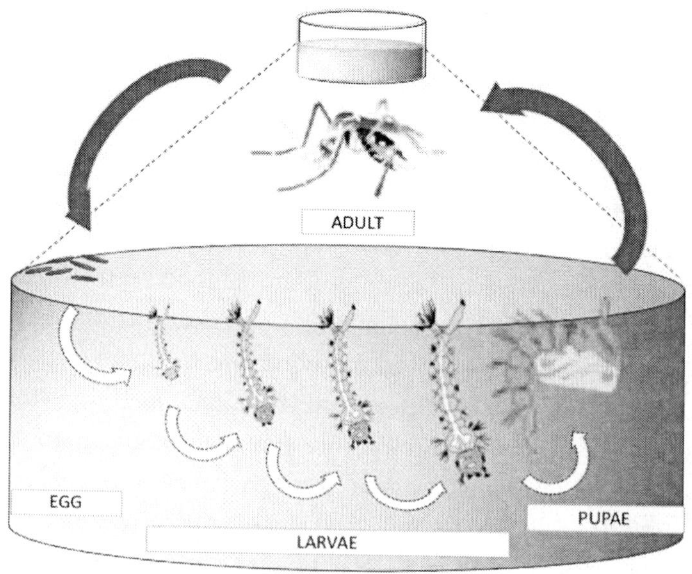

Figure 1. Life stages of *Aedes aegypti* mosquito.

1.2.1. Eggs

Adult *Aedes aegypti* females lay their eggs on the sides of the water holding containers, just above the waterline. These containers typically have small water volume and the presence of detritus (Garcia-Sanchez, Pinilla et al. 2017). Usually, 100 eggs at a time are laid, but warmer temperatures (22°C and 32°C) cause a decreasing egg-laying time and increasing the egg number (Marinho, Beserra et al. 2016) (Ariani, Smith et al. 2015). Eggs are very hardy; they stick to the walls of a container like glue and can survive drying out for several months (Centers for Disease Control and Prevention. U.S 2020). *Aedes aegypti* eggs have adaptive characteristics of resistance to desiccation (Juliano, O'Meara et al. 2002) which allows them to survive in inhospitable environments as well as favoring their transportation voluntarily or involuntarily to different places (Guagliardo, Morrison et al. 2015). Two ecological factors are important for this stage: warm temperature and water.

1.2.2. Larva

Larvae emerge from mosquito eggs after the water level rises enough to cover the eggs. This means that rainwater or humans adding water to containers with dormant eggs will trigger the larvaê to emerge. *Aedes aegypti* larvae need to molt three times before it becomes a pupae in about a week. Some abiotic (e.g., warm temperature, phosphorus, presence of detritus) and biotic factors (e.g., microorganisms such as Cyanobacteria) (Garcia-Sanchez, Pinilla et al. 2017) contribute for a rapid development of larvae. Detritus is a determining element in the provision of energy for immature-stages of *Aedes aegypti* because it is the primary source of essential nutrients that support the trophic chain of these habitats (Garcia-Sanchez, Pinilla et al. 2017). The presence of organic matter (biotic factor) seems to accelerate the developmental time from the first larval stage from larval stage to the adult mosquito. (Tun-Lin, Burkot et al. 2000).

1.2.3. Pupae

Pupae will develop for about 2 days in the water until the body of the newly formed adult flying mosquitoes emerges from and leaves the pupal skin. In contrast to the environment of the larval stage, low level of nutrients and oxygen may be present and, probably, in response to ecdysteroid hormone. (Telang, Frame et al. 2007).

1.2.4. Adult

The adult emerged, *Aedes aegypti* are a medium-sized mosquito approximately 4-7 mm in length Adults have white scales on the dorsal surface of their thorax and a dark brown to black abdomen with white scales. Tarsal segments of the hind legs have white basal bands that form what appear to be stripes (Clemons, Haugen et al. 2010). Newly emerged adult *Aedes aegypti* will rest inside houses, often in quiet, dark places like closets or clothes racks. Another important characteristic of this life stage is that adults often will live their entire life in a single house or its neighbor, and seldom fly beyond 100 meters from their initial resting site. Male mosquitoes feed exclusively on nectar from flowers, while female mosquitoes require to feed on a blood source for protein for eggs production. *Aedes aegypti* is mammophilic (i.e., feeds on mammals) and females will readily feed on human blood. After feeding, female mosquitoes will look for water sources to lay their eggs. Although adult populations of *Aedes aegypti* tend to be low, for example, is common to find 10 female adult per house (Morrison, Sihuincha et al. 2006) they are highly efficient to transmit arbovirus due to their preference for an indoor habitat, frequent human biting and their vectorial competence.

2. Control of *Aedes aegypti*

The control of populations of *Aedes aegypti* is performed using strategies such as integrated control of environmental, chemical, and biological management. Environmental management is the most effective for prevention or reduction of immature mosquitoes and the contact between humans, vectors, and pathogen. Environmental management is based on the destruction, alteration, or removal of potential water sources used by the larvae that can produce the greatest amounts of *Aedes aegypti* in the community. These sources can be artificial habitats such as containers and/or natural habitats, like tree holes. These activities are carried out simultaneously with health educational programs, using communication strategies that encourages community participation in planning, implementation and evaluation of the programs for management of containers, for example.

Community participation is key to this strategy, and several clinical trials based on this approach, have shown an increase in the effectiveness of vector control by governments (Andersson, Arostegui et al. 2017). Unfortunately, it is very common that continuity and sustainability of these strategies have not yet been incorporated into vector control policies adopted by the governments. Therefore, environmental management is underutilized. On the other hand, the most widely used method for vector control is chemical control due to its short-term regular effectiveness against larvae and adult populations of *Aedes aegypti*. However, pesticide resistance has become a serious health thread in different latitudes. For this reason, other forms of control are being developed including biological control using Wolbachia, and genetic control through genetically modified mosquitoes. Results from experimental studies are promising but the long-term environment results of their efficacy, the ecological impact and the ethical implications need to be better defined.

2.1. Chemical Control

2.1.1. Insecticides

In several countries of The Americas, chemical control of *Aedes aegypti* has been the most important tool, whether in interepidemic or epidemic periods of time. During interepidemic periods, the "focal treatment" or use of insecticides (organophosphate) is widely used and placed in potential breeding (eggs and larvae) sites of the vector, such as containers or water reservoirs. But during epidemic periods, in addition to the use of larvicides, large-scale indoor ultra-low-volume space spraying of insecticides and the removal of unusable materials or management of the environment have been most widely used. In Table 1, insecticides that had been used and that are used are listed.

Table 1. Insecticides for *Aedes aegypti* control

Chemical group	Insecticide
Organochlorine	Dichloro-Diphenyl-Trichloro-ethane (DDT)
Organophosphate	Temephos
	Malathion
	Methyl-pyrimifos
	Fenitrothion
	Chlorpyrifos
Carbamates	Propoxur
	Bendiocarb
Pyrethroids	Deltamethrin
	Lambda-cyhalothrin
	Cypermethrin
	Cyfluthrin
	Permethrin
	Alpha-cypermethrin
	Pyrethrum
	Bifenthrin
	D-phenothrin
	Z-Cypermethrin

2.1.2. Criteria and Considerations for Application of Chemical Control

In regions with historical presence of the vector, the unusual increase of dengue fever cases has led to a consideration of large-scale indoor ultra-low-volume space spraying of insecticides and environmental management in the affected areas. In order to define the timing and the area of control intervention, a key first criteria used by the control agencies has mainly relied on the *Aedes aegypti* larval index. This has included the *Aedes* index, Breteau index or the positivity rate of ovitrap. Currently, it is known that the larval indices (*Aedes* index, Breteau index) do not have a good correlation with the incidence of dengue disease, unlike the adult index (Morrison, Minnick et al. 2010). It has a high implication on vector control, and why, in Iquitos, Peru, vector control agencies started using for their entomological surveillance, a light and simple device to capture adult mosquitoes, known as the "Prokopak" (Vazquez-Prokopec, Galvin et al. 2009). A second criteria, is the investigation of the reported dengue cases in order to track the probable site if infection (epidemiological component); however, studies of transmission dynamics of the dengue virus have shown the importance of other place different to the workplace or residence as a probable place of infection. The high mobility of infected individuals (Stoddard, Morrison et al. 2009) in some areas endemic for dengue virus, increases the risk of infection into the "space of activity." This is defined as the most important places where the dengue virus-infected person has been in contact with other people in the last week. This concept has enormous importance for chemical control increasing our field investigation and level of suspicion to include other scenarios as probable sites of infection beyond only the place of residence and place of work.

High urban growth of sprawling urban areas, where there is great presence of the *Aedes aegypti*, not only has led to a favorable ecological environment, but also has complicated entomological surveillance and planning of vector control due to significant variations in terms of housing's features (subdivisions), codification and access.

The utilization of Geographic Information System (GIS) in the epidemiological and entomological surveillances is helping with a data visualization and calculation of risky areas. This may help facilitate the best use of vector control resources. However, one limitation, especially, is the fact that this system relies on epidemiological data. In most developing countries, the epidemiological surveillance of dengue, Zika or chikungunya cases, is passive. This surveillance is characterized by a low sensitivity. Also, a large proportion of asymptomatic carriers of Zika and dengue (Morrison, Minnick et al. 2010) (Chastel 2012) will not be detected. In the case of entomological surveillance, the data needs to be standard (households coding) and updated. An additional problem is the finding of abandoned houses and reluctant households to the indoor ultra-low-volume space spraying or focal treatment (larvicides).

2.1.3. Frequent Types of Chemical Control

2.1.3.1. Focal Treatment or Larviciding

The focal treatment or control of breeding sites physically or chemically, by larviciding, has been one of the most important strategies for prevention and control of the larval stage of *Aedes aegypti* in the last 20 years. This strategy has required the preparation and distribution of the insecticide inside homes and community areas. Usually, it has been developed by health personnel during interepidemic periods using larvicides such as temephos. On the other hand, physical removal of breeding sites has been a task carried out by health personnel mainly during epidemic periods of dengue and has been denominated as "removal of unusable materials" (management of the environment). In that scenario, community mobilization has been promoted shortly and successfully, by the control vector agencies using the press and radio. Community mobilization during inter-epidemic periods has been low as well, as has the support from government programs that promote this. Moreover, resistance to temephos has been reported in many countries, and therefore, in some countries such as Brazil and Peru, the use of temephos was stopped, and different organophosphorus such as pyriproxyphene was considered (Brasil 2014).

2.1.3.2. Indoor Space Spraying Using Ultra-Low Volume

Indoor space spraying using ultra-low volume is one of the first actions to control dengue epidemics in several countries of Central and South America. This is performed through three applications separated by 3 to 5 days in the affected areas. Entomological surveillance tolls are used to measure the effectiveness and efficiency. The vector control using indoor space spraying would be more effective if standard operational procedures are followed, with strict team supervision during the spraying until 85% of the houses have been treated.

Problems observed in developing countries, when using indoor space spraying of ultra-low volume, for example, in Iquitos, Peru, are:

- The number of personnel required to carry out large-scale indoor space spraying using ultra-low volume entails hiring new personnel with inadequate training and technical profile for this function, which increases the risk of programmatic or technical errors.
- Field tests of insecticide resistance had not been a priority activity for the decision of purchase of insecticides by regional or central governments. For example, resistance to temephos had already been reported but the control vector agencies were still purchasing and applying it.
- Regulation to reduce the number of households reluctant to surveillance or vector control activities. There is a need for laws and/or regulations that allow legal or economic action against those who do not allow the deployment of monitoring or vector control activities. For example, in the case of Peru, the law of protection of the inviolability of the domicile looks to be superior to any other law; however, there is another public health law that enables action (Vilcarromero, Casanova et al. 2015). Therefore, laws must allow the deployment of control activities at any stage and under any scenario.
- On the other hand, perception and comfort of the population when deploying vector control activities should be taken into consideration, to educate and assess the impact of these activities. For example, in Iquitos, Peru, a study about population perception (Paz-Soldan, Morrison et al. 2015) found that the concerned population had the perception that "spatial fumigation was useless" and, although those comments were interpreted as probable confusion of the population due to the use of insecticides targeting *A. aegypti* but no other mosquitoes. This confusion was due to increased populations of other mosquitoes, such as Culex, which had increased after fumigation. However, it was some few weeks

later known that these were signs of lack of efficacy of the chemical control measures due to several factors including their resistance to temephos. Furthermore, in developing countries, the affected population is of low income and cannot afford to stop working even to meet with the staff doing the fumigation. Alternatively, an advance notice of the most convenient times should be obtained for the houses that previously were found to be closed. It clearly highlights the importance to have a communication plan with the public and provide options to improve access to homes for vector control and education.

- Resistance to the insecticides used to control the adult stage has also been found for a range of pyrethroid insecticides and organophosphorus, like malathion in some countries of Southeast Asian (Amelia-Yap, Chen et al. 2018). Similarly, it was detected resistance to some pyrethroids such as lambda-cyhalothrin and possible resistance to permethrin in the region of Jazan in Southwestern Saudi Arabia (Dafalla, Alsheikh et al. 2019). There is also high frequency of reported resistance to Deltamethrin in vector populations from Latin America and The Caribbean (Guedes, Beins et al. 2020)
- Vector control measures need community mobilization, so education in mosquito prevention needs to be performed in a long-term period.
- A multi sectorial approach is needed in the fight against *Ae. aegypti*-transmitted arboviruses. Ecological factors for the mosquito development such as: poor sanitation conditions, lack of potable water, no sewage, and no proper public garbage collection service requires the participation of social actors, public sector and private institutions.

2.1.3.3. Other Intradomicile Methods: Indoor Residual Spraying (and Insecticide-Treated Materials

Intradomicile space spraying (indoor space spraying using ultra-low volume) and indoor residual spraying target the endophilic adult *Aedes* mosquitoes that bite and rest indoors. Insecticide-treated materials (e.g., curtains) entails the coating of walls and surfaces of the entire house with a residual insecticide. The use of intradomicile space fumigation targets indoor flying insects with less residual effect. Trial results from intradomicile application of indoor residual spraying and insecticide-treated materials are encouraging because they reduce *Aedes aegypti* populations and human infections (Scott TW Morrison AC 2010). Moreover, intradomicile adulticide interventions do more than decrease adult mosquito density. They reduce vector lifespan, which has a greater overall impact on virus transmission than reducing vector density alone. In addition, intradomicile control will decrease other vector borne diseases and insect pest problems (Gunning, Okamoto et al. 2018). Even if adult mosquito populations recover quickly, indoor application of insecticides can interrupt epidemic virus transmission, resulting in a noticeable impact on disease (Scott TW Morrison AC 2010). However, a recent meta-analysis (Bowman, Donegan et al. 2016) concluded that there is a need of more empirical evidence supporting the potential utility of indoor residual spraying for dengue prevention, since it was based on only two studies.

2.1.3.4. Outdoor Spatial Spraying Using Ultra-Low Volume

Outdoor spatial spraying using ultra-low volume fogging has been one of the most common control vector practices used by local governments to target adult mosquito vectors. The insecticides (typically synthetic pyrethroids or organophosphates) are applied to outdoor areas using truck-mounted equipment or backpack sprayers (Tidwell, Williams et al. 1994). Evidence for the efficacy of outdoor spatial spraying for dengue prevention and control is limited (Bowman, Donegan et al. 2016) (Reyes-Castro, Castro-Luque et al. 2017) and one important limitation is its poor spray penetrance to accessible spaces in and around homes as well as the use of chemicals with no residual activity (Reiter P 2014). In addition, repeated use of particular insecticides has been associated with the development of resistance in the vector (Marcombe, Darriet et al. 2011) (Reyes-Castro, Castro-Luque et al. 2017). This is the main reason why this method is not recommended.

2.2. Genetic and Biological Control

Two new approaches have been tested during the recent years in field trials. The first is based on the release of genetically modified male mosquitoes that carry a dominant lethal gene. Oxitec (Abingdon, UK) has developed genetically modified (GM) strains of *A. aegypti* via the release of insects carrying a dominant lethal (RIDL) strategy (Qsim, Ashfaq et al. 2017). This technique using RIDL mosquitoes is considered to be ecologically friendly and specific and has been shown to be able to suppress mosquito populations but requires the continuous release of large numbers of transgenic mosquitoes for several months (Dorigatti, McCormack et al. 2018). However, in the absence of a gene drive system, questions remain about the sustainability of this intervention, as the migration of unmodified mosquitoes from neighboring areas following local suppression of a mosquito population requires releases to be continued indefinitely, albeit at lower levels than during initial suppression (Carvalho, McKemey et al. 2015), (Huang, Higgs et al. 2017). Other concerns are referred to their final cost, efficacy in different settings, and patents questions attributed to a genetic product (Meghani and Boete 2018).

The second approach involves the release of mosquitoes transinfected with the vertically transmitted intracellular bacterium *Wolbachia*. (See Figure 2). This endosymbiotic bacterium has demonstrated to interfere with the arboviral infections in *Aedes aegypti* and can be used for the control of arboviral infections in one of two strategies: the reduction of vectors' reproductive capacity and the induction of resistance to RNA viruses (Qsim, Ashfaq et al. 2017). Wolbachia can induce a form of sterility known as Cytoplasmic Incompatibility (CI), in which embryos from uninfected females fertilized by sperm from infected males fail to develop. Infected males are therefore sterile when mated with uninfected females, though fertile when mating with infected females (Alphey 2014). With the possibility of Wolbachia providing refractoriness for several arboviral infections and a Gene Drive System, make this as a most promising novel intervention to control dengue and other arboviruses.

However, effectiveness in large trials is still necessary to calculate (Dorigatti, McCormack et al. 2018).

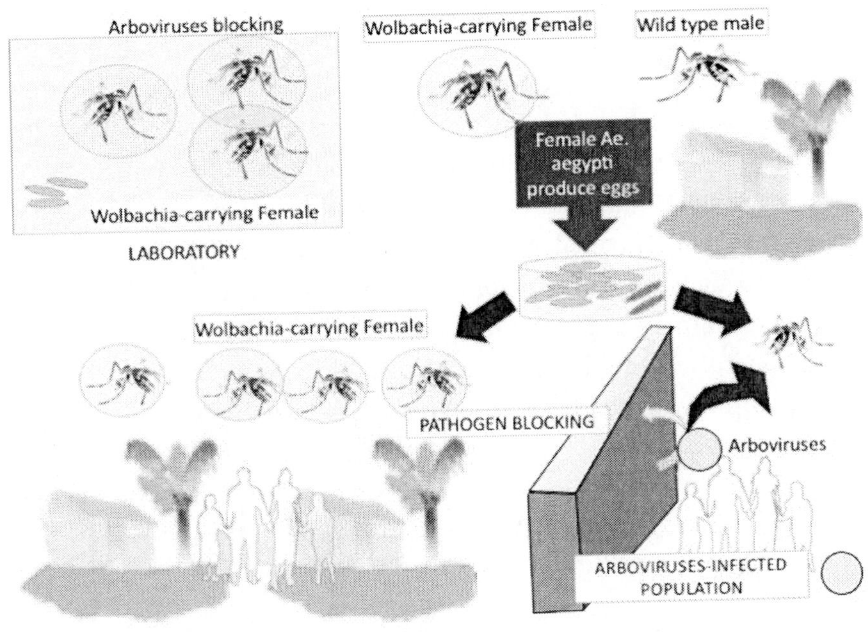

Figure 2. Vector control using Wolbachia.

3. Arboviral Transmission

Aedes aegypti mosquito, is responsible for transmission of several important viral infections, including yellow fever, dengue, chikungunya, and Zika virus. Yellow fever is currently endemic in several countries of Africa and America and is responsible for approximately 200,000 cases of disease and 30,000 deaths each year (Centers for Disease Control and Prevention. U.S 2018). Dengue is the world's most important arboviral disease in terms of number of people affected. Over the past 50 years, incidence increased 30-fold: there were approximately 390 million infections in 2010 (Ebi and Nealon 2016). The total number of disability-adjusted life years (DALYs) lost due to dengue was 1.14 million in 2013 alone (Stanaway, Shepard et al. 2016). Chikungunya and Zika virus were usually involved in localized outbreaks, however in recent years, they were involved in epidemics in Central and South America (Borchering, Huang et al. 2019) (Mogami, de Almeida Vieira et al. 2017). In these epidemics, the neurotropism of Zika virus was demonstrated causing Guillain Barre syndrome and neurological impairment of fetus and newborns (Capasso, Ompad et al. 2019) (Einspieler, Utsch et al. 2019). Chikungunya, is characterized by a high proportion of symptomatic cases and by causing a significant long-term morbidity in the form of debilitating arthralgia (Chang, Encinales et al. 2018) (van Aalst, Nelen et al. 2017). In addition to the medically important arboviruses mentioned above, these *Aedes* mosquito species are also the potential vectors of several other endemic viral infections such as West Nile fever and Mayaro virus infection (Long, Ziegler et al. 2011) (Daep, Munoz-Jordan et al. 2014).

3.1. Arbovirus-Mosquito Vector-Host Interactions and Transmission

Only female *Aedes aegypti* female feed on blood for egg production. Since *Aedes aegypti* is mammophilic (i.e., feeds on mammals), females will readily feed on human blood. They are "day-biters" and tend to bite during the daytime when people are engaged in their daily activities. Three important characteristics in the arbovirus transmission by *Aedes aegypti* are:

1. The high adaptation to the human environment (indoor). This mosquito species live, bites and will lay eggs close to humans. A rapid urban grow has created the optimal ecological conditions for the *Aedes aegypti* development. Humans are the main viral reservoir for dengue, chikungunya, and Zika virus outbreaks or epidemics. Viral amplification occurs when uninfected mosquitoes bite viremic humans and viremic mosquitoes bite uninfected humans. This increases the virus levels in both humans and

mosquitoes leading to increased human cases and public health risk.

2. Frequent human biting. An *Aedes aegypti* female bites several times per day seeking blood. This characteristic makes this mosquito species, an efficient arboviruses vector, despite the low population densities in the house or neighborhood. This explains how an arbovirus outbreak can start with such low entomological thresholds. This property also increase their fitness and lets this female adult mosquito specie to lay more eggs. (Morrison, Gray et al. 2004).
3. Vector competence of *Aedes aegypti* for arboviruses. *Aedes aegypti* vector competence studies have shown this species to be variable in the vector competency. This makes it difficult to reach an simple conclusion, however, the following can be inferred: Complete susceptibility to infection has been observed for Zika virus (ZIKV), dengue virus (DENV) and chikungunya virus (CHIKV), but not yellow fever viruses (YFV); (2) the dose of virus used is directly correlated to the rate of infection; (3) Brazilian populations of mosquito are particularly susceptible to DENV-2 infections; (4) the Asian lineage of ZIKV is less infective to *Ae. aegypti* populations from the American continent than is the African ZIKV lineage; (5) virus adaptation to different species of mosquitoes has been demonstrated with CHIKV (Souza-Neto, Powell et al. 2019) (Bonica, Goenaga et al. 2019).

CONCLUSION

Rapid urban growth, high population density has favored the shortage of drinking water, the mismanagement of solid waste, lack of siphons, etc. This has favored a great interaction and the best ecological conditions for the development of the stages of *Aedes aegypti*. Some biological characteristics of this vector, such as its high frequency in bites, have made this mosquito a very efficient vector in the transmission of arboviruses such as dengue, Zika and chikungunya. The observed epidemics are the result of this interaction. During these epidemics, the use of insecticides has been the most important element used by the governments of the affected countries. Unfortunately, this extensive and varied use from country to country has generated resistance to insecticides. New forms of control have been developed, using engineered mosquitoes, in which their genes have been modified, to decrease their fertility, and ultimately to block arbovirus transmission. Its effectiveness is still to be assessed, but experimental studies have been promising in certain scenarios. While ethical and cost aspects will be topics for future discussions. The use of biological agents, such as *Wolbachia* has been widely evaluated, and seems to be successful in localized scenarios. It will be important to evaluate its effectiveness. Meanwhile, there is no single form of control for *Aedes aegypti* and more than one control strategy should be considered according to the characteristics of each affected region and considering the empowerment of the population.

REFERENCES

(1970). "[Erradication of malaria]." *Bol Oficina Sanit Panam* 69(2): 160-166.

Alphey, L. (2014). "Genetic control of mosquitoes." *Annu Rev Entomol* 59: 205-224.

Amelia-Yap, Z. H., C. D. Chen, M. Sofian-Azirun and V. L. Low (2018). "Pyrethroid resistance in the dengue vector *Aedes aegypti* in Southeast Asia: present situation and prospects for management." *Parasit Vectors* 11(1): 332.

Andersson, N., J. Arostegui, E. Nava-Aguilera, E. Harris and R. J. Ledogar (2017). "Camino Verde (The Green Way): evidence-based community mobilisation for dengue control in Nicaragua and Mexico: feasibility study and study protocol for a randomised controlled trial." *BMC Public Health* 17(Suppl 1): 407.

Ariani, C. V., S. C. Smith, J. Osei-Poku, K. Short, P. Juneja and F. M. Jiggins (2015). "Environmental and genetic factors determine whether the mosquito *Aedes aegypti* lays eggs without a blood meal." *Am J Trop Med Hyg* 92(4): 715-721.

Bonica, M. B., S. Goenaga, M. L. Martin, M. Feroci, V. Luppo, E. Muttis, C. Fabbri, M. A. Morales, D. Enria, M. V. Micieli and S. Levis (2019). "Vector competence of *Aedes aegypti* for different strains of Zika virus in Argentina." *PLoS Negl Trop Dis* 13(6): e0007433.

Borchering, R. K., A. T. Huang, Y. T.-R. L. Mier, D. P. Rojas, I. Rodriguez-Barraquer, L. C. Katzelnick, S. D. Martinez, G. D. King, S. C. Cinkovich, J. Lessler and D. A. T. Cummings (2019). "Impacts of Zika emergence in Latin America on endemic dengue transmission." *Nat Commun* 10(1): 5730.

Bowman, L. R., S. Donegan and P. J. McCall (2016). "Is Dengue Vector Control Deficient in Effectiveness or Evidence?: Systematic Review and Meta-analysis." *PLoS Negl Trop Dis* 10(3): e0004551.

Brasil, M. d. S. S. d. V. e. S. L. (2014). "*Larvicidas*." Retrieved March 1th, 2020, from https://www.saude.gov.br/saude-de-a-z/controle-de-vetores-inseticidas-e-larvicidas/larvicidas.

Capasso, A., D. C. Ompad, D. L. Vieira, A. Wilder-Smith and Y. Tozan (2019). "Incidence of Guillain-Barre Syndrome (GBS) in Latin America and the Caribbean before and during the 2015-2016 Zika virus epidemic: A systematic review and meta-analysis." *PLoS Negl Trop Dis* 13(8): e0007622.

Carvalho, D. O., A. R. McKemey, L. Garziera, R. Lacroix, C. A. Donnelly, L. Alphey, A. Malavasi and M. L. Capurro (2015). "Suppression of a Field Population of *Aedes aegypti* in Brazil by Sustained Release of Transgenic Male Mosquitoes." *PLoS Negl Trop Dis* 9(7): e0003864.

Centers for Disease Control and Prevention. U.S. (2018). *Yellow Fever*. Retrieved March, 2th, 2020, from https://www.cdc.gov/globalhealth/newsroom/topics/yellowfever/index.html.

Centers for Disease Control and Prevention. U.S. (2020). *Mosquito life cycle. Aedes aegypti*. Retrieved March 1th, 2020, from https://www.cdc.gov/zika/pdfs/MosquitoLifecycle.pdf.

Chang, A. Y., L. Encinales, A. Porras, N. Pacheco, S. P. Reid, K. A. O. Martins, S. Pacheco, E. Bravo, M. Navarno, A. Rico Mendoza, R. Amdur, P. Kamalapathy, G. S. Firestein, J. M. Bethony and G. L. Simon (2018). "Frequency of Chronic Joint Pain Following Chikungunya Virus Infection: A Colombian Cohort Study." *Arthritis Rheumatol* 70(4): 578-584.

Chastel, C. (2012). "Eventual role of asymptomatic cases of dengue for the introduction and spread of dengue viruses in non-endemic regions." *Front Physiol* 3: 70.

Clemons, A., M. Haugen, E. Flannery, M. Tomchaney, K. Kast, C. Jacowski, C. Le, A. Mori, W. Simanton Holland, J. Sarro, D. W. Severson and M. Duman-Scheel (2010). "*Aedes aegypti*: an emerging model for vector mosquito development." *Cold Spring Harb Protoc* 2010(10): pdb emo141.

Daep, C. A., J. L. Munoz-Jordan and E. A. Eugenin (2014). "Flaviviruses, an expanding threat in public health: focus on dengue, West Nile, and Japanese encephalitis virus." *J Neurovirol* 20(6): 539-560.

Dafalla, O., A. Alsheikh, W. Mohammed, K. Shrwani, F. Alsheikh, Y. Hobani and E. Noureldin (2019). "Knockdown resistance mutations contributing to pyrethroid resistance in *Aedes aegypti* population, Saudi Arabia." *East Mediterr Health J* 25(12): 905-913.

Dorigatti, I., C. McCormack, G. Nedjati-Gilani and N. M. Ferguson (2018). "Using Wolbachia for Dengue Control: Insights from Modelling." *Trends Parasitol* 34(2): 102-113.

Ebi, K. L. and J. Nealon (2016). "Dengue in a changing climate." *Environ Res* 151: 115-123.

Einspieler, C., F. Utsch, P. Brasil, C. Y. Panvequio Aizawa, C. Peyton, R. Hydee Hasue, F. Francoso Genovesi, L. Damasceno, M. E. Moreira, K. Adachi, P. B. Marschik, K. Nielsen-Saines and G. M. Z. W. Group (2019). "Association of Infants Exposed to Prenatal Zika Virus Infection With Their Clinical, Neurologic, and Developmental Status Evaluated via the General Movement Assessment Tool." *JAMA Netw Open* 2(1): e187235.

Eisen, L., A. J. Monaghan, S. Lozano-Fuentes, D. F. Steinhoff, M. H. Hayden and P. E. Bieringer (2014). "The impact of temperature on the bionomics of *Aedes* (*Stegomyia*) *aegypti*, with special reference to the cool geographic range margins." *J Med Entomol* 51(3): 496-516.

Garcia-Sanchez, D. C., G. A. Pinilla and J. Quintero (2017). "Ecological characterization of *Aedes aegypti* larval habitats (Diptera: Culicidae) in artificial water containers in Girardot, Colombia." *J Vector Ecol* 42(2): 289-297.

Guagliardo, S. A., A. C. Morrison, J. L. Barboza, E. Requena, H. Astete, G. Vazquez-Prokopec and U. Kitron (2015). "River boats contribute to the regional spread of the dengue vector *Aedes aegypti* in the Peruvian Amazon." *PLoS Negl Trop Dis* 9(4): e0003648.

Guedes, R. N. C., K. Beins, D. Navarro Costa, G. E. Coelho and H. Bezerra (2020). "Patterns of insecticide resistance in *Aedes aegypti*: meta-analyses of surveys in Latin America and the Caribbean." *Pest Manag Sci*.

Gunning, C. E., K. W. Okamoto, H. Astete, G. M. Vasquez, E. Erhardt, C. Del Aguila, R. Pinedo, R. Cardenas, C. Pacheco, E. Chalco, H. Rodriguez-Ferruci, T. W. Scott, A. L. Lloyd, F. Gould and A. C. Morrison (2018). "Efficacy of *Aedes aegypti* control by indoor Ultra Low Volume (ULV) insecticide spraying in Iquitos, Peru." *PLoS Negl Trop Dis* 12(4): e0006378.

Holstein, M. (1967). "Dynamics of *Aedes aegypti* distribution, density and seasonal prevalence in the Mediterranean area." *Bull World Health Organ* 36(4): 541-543.

Huang, Y. S., S. Higgs and D. L. Vanlandingham (2017). "Biological Control Strategies for Mosquito Vectors of Arboviruses." *Insects* 8(1).

Juliano, S. A., G. F. O'Meara, J. R. Morrill and M. M. Cutwa (2002). "Desiccation and thermal tolerance of eggs and the coexistence of competing mosquitoes." *Oecologia* 130(3): 458-469.

Long, K. C., S. A. Ziegler, S. Thangamani, N. L. Hausser, T. J. Kochel, S. Higgs and R. B. Tesh (2011). "Experimental transmission of Mayaro virus by *Aedes aegypti*." *Am J Trop Med Hyg* 85(4): 750-757.

Lwande, O. W., V. Obanda, G. Bucht, G. Mosomtai, V. Otieno, C. Ahlm and M. Evander (2015). "Global emergence of Alphaviruses that cause arthritis in humans." *Infect Ecol Epidemiol* 5: 29853.

Marcombe, S., F. Darriet, M. Tolosa, P. Agnew, S. Duchon, M. Etienne, M. M. Yp Tcha, F. Chandre, V. Corbel and A. Yebakima (2011). "Pyrethroid resistance reduces the efficacy of space sprays for dengue control on the island of Martinique (Caribbean)." *PLoS Negl Trop Dis* 5(6): e1202.

Marinho, R. A., E. B. Beserra, M. A. Bezerra-Gusmao, S. Porto Vde, R. A. Olinda and C. A. Dos Santos (2016). "Effects of temperature on the life cycle, expansion, and dispersion of *Aedes aegypti* (Diptera: Culicidae) in three cities in Paraiba, Brazil." *J Vector Ecol* 41(1): 1-10.

Meghani, Z. and C. Boete (2018). "Genetically engineered mosquitoes, Zika and other arboviruses, community engagement, costs, and patents: Ethical issues." *PLoS Negl Trop Dis* 12(7): e0006501.

Mogami, R., A. de Almeida Vieira, E. A. Junqueira Filho and A. J. Lopes (2017). "Chikungunya fever outbreak in Rio de Janeiro, Brazil: Ultrasonographic aspects of musculoskeletal complications." *J Clin Ultrasound* 45(1): 43-44.

Morrison, A. C., K. Gray, A. Getis, H. Astete, M. Sihuincha, D. Focks, D. Watts, J. D. Stancil, J. G. Olson, P. Blair and T. W. Scott (2004). "Temporal and geographic patterns of *Aedes aegypti* (Diptera: Culicidae) production in Iquitos, Peru." *J Med Entomol* 41(6): 1123-1142.

Morrison, A. C., S. L. Minnick, C. Rocha, B. M. Forshey, S. T. Stoddard, A. Getis, D. A. Focks, K. L. Russell, J. G. Olson, P. J. Blair, D. M. Watts, M. Sihuincha, T. W. Scott and T. J. Kochel (2010). "Epidemiology of dengue virus in Iquitos, Peru 1999 to 2005: interepidemic and epidemic patterns of transmission." *PLoS Negl Trop Dis* 4(5): e670.

Morrison, A. C., M. Sihuincha, J. D. Stancil, E. Zamora, H. Astete, J. G. Olson, C. Vidal-Ore and T. W. Scott (2006). "*Aedes aegypti* (Diptera: Culicidae) production from non-residential sites in the Amazonian city of Iquitos, Peru." *Ann Trop Med Parasitol* 100 Suppl 1: S73-S86.

Paz-Soldan, V. A., A. C. Morrison, J. J. Cordova Lopez, A. Lenhart, T. W. Scott, J. P. Elder, M. Sihuincha, T. J. Kochel, E. S. Halsey, H. Astete and P. J. McCall (2015). "Dengue Knowledge and Preventive Practices in Iquitos, Peru." *Am J Trop Med Hyg* 93(6): 1330-1337.

Powell, J. R. and W. J. Tabachnick (2013). "History of domestication and spread of *Aedes aegypti*--a review." *Mem Inst Oswaldo Cruz* 108 Suppl 1: 11-17.

Qsim, M., U. A. Ashfaq, M. Z. Yousaf, M. S. Masoud, I. Rasul, N. Noor and A. Hussain (2017). "Genetically Modified *Aedes aegypti* to Control Dengue: A Review." *Crit Rev Eukaryot Gene Expr* 27(4): 331-340.

Reiter P (2014). Surveillance and Control of Dengue vectors. O. E. In: Gubler DJ, Vasudevan S, Farrar J, editors. *Dengue and Dengue Hemorrhagic Fever*, Wallingford: CAB International.

Reyes-Castro, P. A., L. Castro-Luque, R. Diaz-Caravantes, K. R. Walker, M. H. Hayden and K. C. Ernst (2017). "Outdoor spatial spraying against dengue: A false sense of security among inhabitants of Hermosillo, Mexico." *PLoS Negl Trop Dis* 11(5): e0005611.

Scott TW Morrison AC (2010). Vector Dynamics and Transmission of Dengue Virus: Implications for Dengue Surveillance and Prevention Strategies. *Dengue Virus. Current Topics in Microbiology and Immunology*. I. R. A, Springer, Berlin, Heidelberg: 115-128.

Smith, C. E. (1956). "The history of dengue in tropical Asia and its probable relationship to the mosquito *Aedes aegypti*." *J Trop Med Hyg* 59(10): 243-251.

Souza-Neto, J. A., J. R. Powell and M. Bonizzoni (2019). "*Aedes aegypti* vector competence studies: A review." *Infect Genet Evol* 67: 191-209.

Stanaway, J. D., D. S. Shepard, E. A. Undurraga, Y. A. Halasa, L. E. Coffeng, O. J. Brady, S. I. Hay, N. Bedi, I. M. Bensenor, C. A. Castaneda-Orjuela, T. W. Chuang, K. B. Gibney, Z. A. Memish, A. Rafay, K. N. Ukwaja, N. Yonemoto and C. J. L. Murray (2016). "The global burden of dengue: an analysis from the Global Burden of Disease Study 2013." *Lancet Infect Dis* 16(6): 712-723.

Stoddard, S. T., A. C. Morrison, G. M. Vazquez-Prokopec, V. Paz Soldan, T. J. Kochel, U. Kitron, J. P. Elder and T. W. Scott (2009). "The role of human movement in the transmission of vector-borne pathogens." *PLoS Negl Trop Dis* 3(7): e481.

Telang, A., L. Frame and M. R. Brown (2007). "Larval feeding duration affects ecdysteroid levels and nutritional reserves regulating pupal commitment in the yellow fever mosquito *Aedes aegypti* (Diptera: Culicidae)." *J Exp Biol* 210(Pt 5): 854-864.

Tidwell, M. A., D. C. Williams, T. A. Gwinn, C. J. Pena, S. H. Tedders, G. E. Gonzalvez and Y. Mekuria (1994). "Emergency control of *Aedes aegypti* in the Dominican Republic using the Scorpion 20 ULV forced-air generator." *J Am Mosq Control Assoc* 10(3): 403-406.

Tun-Lin, W., T. R. Burkot and B. H. Kay (2000). "Effects of temperature and larval diet on development rates and survival of the dengue vector *Aedes aegypti* in north Queensland, Australia." *Med Vet Entomol* 14(1): 31-37.

van Aalst, M., C. M. Nelen, A. Goorhuis, C. Stijnis and M. P. Grobusch (2017). "Long-term sequelae of chikungunya virus disease: A systematic review." *Travel Med Infect Dis* 15: 8-22.

Vilcarromero, S., W. Casanova, J. S. Ampuero, C. Ramal-Asayag, C. Siles, G. Diaz, S. Durand, J. C. Celis-Salinas, H. Astete, P. Rojas, G. Vasquez-La Torre, J. Marin, I. Bazan, Y. Alegre, A. C. Morrison and H. Rodriguez-Ferrucci (2015). "[Lessons learned in the control of *Aedes aegypti* to address dengue and the emergency of chikungunya in Iquitos, Peru]." *Rev Peru Med Exp Salud Publica* 32(1): 172-178.

Biographical Sketches

Stalin Vilcarromero

Affiliation: Stony Brook University, New York.

Education: MD. DTM&H, MPH (c).

Research and Professional Experience: 12 years' experience in arboviruses transmitted disease research (dengue, Zika and Yellow Fever) in the Peruvian Amazon, 3 years' experience in vector borne disease research in Eastern Suffolk County, New York.

Publications from the Last 3 Years:

Brazil burning! What is the potential impact of the Amazon wildfires on vector-borne and zoonotic emerging diseases? - A statement from an international experts meeting. Bonilla-Aldana, D. K, Suárez J. A, Franco-Paredes C, Vilcarromero S, Mattar S, Gómez-Marín, J. E, Villamil-Gómez W. E, Ruíz-Sáenz J, Cardona-Ospina, J. A, Idarraga-Bedoya, S. E., García-Bustos, J. J., Jimenez-Posada, E. V. y Rodríguez-Morales, A. *J. Travel Med Infect Dis*. 2019 Sep - Oct;31: 101474.

Citomegalovirus disease Post-heart trasplant resistant to Ganciclovir: a case report in a National Insitute. Christian Rojas, Gabriel De la Cruz-Kub, Bryan Valcarcel-Valdivia, Stalin Vilcarromero. 2018. ISSN 0036-3634. Accepted to be published in *Revista Peruana de Medicina Experimental y Salud Pública* volumen:35:2018.

Considerations for the definition of co-infection in leptospirosis cases]. Vilcarromero S, Marin J, Casapia M. *Rev Peru Med Exp Salud Publica*. 2019 Apr-Jun;36(2):360-361.

Feasibility of feeding Aedes aegypti*Aedes aegypti* mosquitoes on dengue virus-infected human volunteers for vector competence studies in Iquitos, Peru. Long KC, Sulca J, Ba-zan I, Astete H, Jaba HL, Siles C, Kocher C, Vilcarromero S. et al. (2019) *PLOS Neglect-ed Tropical Diseases* 13(2): e0007116. https://doi.org/10.1371/journal. pntd.000 7116.

Guaroa Virus and Plasmodium vivax Co-Infections, Peruvian Amazon. Siles C, Elson WH, Vilcarromero S, Morrison AC, Hontz RD, Alava F, et al. Guaroa virus and Plasmodium vivax co-infection, Peruvian Amazon. *Emerg Infect Dis*. 2020 Apr.

Risk factors associated with drug use before imprisonment in Peru. Hernández-Vásquez A, Núñez S, Santero M, Grendas L, Huarez B, Vilcarromero S, Casas-Bendezú M, Braun S, Cortés S, Rosselli D. *Rev Esp Sanid Penit* 2018; 20: 11-20.

The implication of reassortant in the diagnosis and epidemiology of Oropuche virus in Peru. Noé Atamari-Anahui, Maycol Suker Ccorahua-Rios, John A. Cabrera-Enríquez, Stalin Vilcarromero. Accepted to be published in *Salud pública Méx*, 2018.

Thomas Iwanejko

Affiliation: Suffolk County Vector Control

Education: MPS – Stony Brook, NY.

Research and Professional Experience: 23 years experience in mosquito control with Suffolk County Vector Control, 4 years as Director of a county mosquito control program.

Professional appointment: Director, Division of Vector Control, Suffolk County Department of Public Works, New York

Publications:

Aquatic insects of New York salt marsh associated with mosquito larval habitat and their potential utility as bioindicators. *J Insect Sci.* 2011;11:172. Rochlin I, Dempsey ME, Iwanejko T, Ninivaggi DV.

Geostatistical evaluation of integrated marsh management impact on mosquito vectors using before-after-control-impact (BACI) design. Int *J Health Geogr*. 2009 Jun 23;8:35. doi: 10.1186/1476-072X-8-35. Rochlin I, Iwanejko T, Dempsey ME, Ninivaggi DV.

Carlos Tong-Rios

Affiliation: Loreto Regional Directorate of Health, Vector Control Unit, Ministry of Health, Iquitos, Peru

Research and Professional Experience: 10 years' experience in vector borne control in the Peruvian Amazon. Also collaborating with many entomological and epidemiological research in this region.

Professional Appointment: Specialist in Vector Control.

Publications from the Last 3 Years:

Continuous Supply of Plasmodium vivax Sporozoites from Colonized *Anopheles darlingi* in the Peruvian Amazon. Moreno M, Tong-Rios C, Orjuela-Sanchez P, Carrasco-Escobar G, Campo B, Gamboa D, Winzeler EA, Vinetz JM. *ACS Infect Dis*. 2018 Apr 13;4(4):541-548.

Developing Plasmodium vivax Resources for Liver Stage Study in the Peruvian Amazon Region. Orjuela-Sanchez P, Villa ZH, Moreno M, Tong-Rios C, Meister S, LaMonte GM, Campo B, Vinetz JM, Winzeler EA. *ACS Infect Dis*. 2018 Apr 13;4(4):531-540.

Dual RNA-seq identifies human mucosal immunity protein Mucin-13 as a hallmark of Plasmodium exoerythrocytic infection. LaMonte GM, Orjuela-Sanchez P, Calla J, Wang LT, Li S, Swann J, Cowell AN, Zou BY, Abdel-Haleem Mohamed AM, Villa Galarce ZH, Moreno M, Tong Rios C, Vinetz JM, Lewis N, Winzeler EA. *Nat Commun*. 2019 Jan 30;10(1):488.

Freddy Alava

Affiliation: Loreto Regional Directorate of Health, Vector Control Unit, Ministry of Health, Iquitos, Peru

Research and Professional Experience: 10 years' experience in vector borne control in the Peruvian Amazon. Also collaborating with many entomological and epidemiological research in this region.

Professional Appointment: Director of the Vector Control Unit.

Publications from the Last 3 Years:

Automated microscopy for routine malaria diagnosis: a field comparison on Giemsa-stained blood films in Peru. Torres K, Bachman CM, Delahunt CB, Alarcon Baldeon J, Alava F, Gamboa Vilela D, Proux S, Mehanian C, McGuire SK, Thompson CM, Ostbye T, Hu L, Jaiswal MS, Hunt VM, Bell D. Malar J. 2018 Sep 25;17(1):339.

Decreasing proportion of Anopheles darlingi biting outdoors between long-lasting insecticidal net distributions in peri-Iquitos, Amazonian Peru. Prussing C, Moreno M, Saavedra MP, Bickersmith SA, Gamboa D, Alava F, Schlichting CD, Emerson KJ, Vinetz JM, Conn JE. Malar J. 2018 Feb 20;17(1):86.

High-accuracy detection of malaria vector larval habitats using drone-based multispectral imagery. Carrasco-Escobar G, Manrique E, Ruiz-Cabrejos J, Saavedra M, Alava F, Bickersmith S, Prussing C, Vinetz JM, Conn JE, Moreno M, Gamboa D. *PLoS Negl Trop Dis*. 2019 Jan 17;13(1):e0007105.

Malaria vector species in Amazonian Peru co-occur in larval habitats but have distinct larval microbial communities. Prussing C, Saavedra MP, Bickersmith SA, Alava F, Guzmán M, Manrique E, Carrasco-Escobar G, Moreno M, Gamboa D, Vinetz JM, Conn JE. *PLoS Negl Trop Dis.* 2019 May 15;13(5):e0007412.

Nyssorhynchus dunhami: bionomics and natural infection by Plasmodium falciparum and P. vivax in the Peruvian Amazon. Prussing C, Bickersmith SA, Moreno M, Saavedra MP, Alava F, Sallum MAM, Gamboa D, Vinetz JM, Conn JE. Mem Inst Oswaldo Cruz. 2018 Dec 3;113(12):e180380.

Scott R. Campbell

Affiliation: Suffolk County Department of Health Services

Education: BS from Muhlenberg College 1984, PhD from Cleveland State University 1990.

Research and Professional Experience: As a public health professional, I have been conducting surveillance and research on arthropod-borne diseases for approximately 29 years.

Professional Appointment: Laboratory Chief of the Arthropod-Borne Disease Laboratory for the Suffolk County Department of Health Service.

Publications from the Last 3 Years:

Ensemble forecast of human West Nile virus cases and mosquito infection rates. Defelice, N. B., E., Little, S. R. Campbell and J. Shaman. 2017. *Nature Communications*. 8, 14592

Epidemiology of Lyme disease among US Veterans in Long Island, New York. *Ticks and Tick-Borne Diseases*, 10:407-411. Psevdos, G, T. Khoo, R. Chow, C. L. Romano, S. Campbell. 2019.

High levels of local inter- and intra-host genetic variation of West Nile virus and evidence of fine-scale evolutionary pressures. Ehrbar, D. J., K. A. Ngo, S. R. Campbell, L. D. Kramer, A. T. Ciota. 2017. *Infection, Genetics and Evolution,* 51:219-226.

Modeling and Surveillance of Reporting Delays of Mosquitoes and Humans Infected with West Nile Virus and Associations With Accuracy of West Nile Virus Forecasts. Defelice, N. B. R. Birger, N. Defelice, A. Gagner, S. R. Campbell, C. Romano, M. Santoriello, J. Henke, J. Wittie, B. Cole, C. Kaiser, J. Shaman. 2019. *Jama Network Open*. 2019;2(4):e193175. doi:10.1001/jamanetworkopen.2019.3175.

Spatiotemporal modeling of ecological and sociological predictors of West Nile virus in Suffolk County, NY, mosquitoes. Myer, M. H., S. R. Campbell, and J. M. Johnston. 2017. *Ecosphere* 8(6):e01854.

Twenty years of surveillance for Eastern equine encephalitis virus in mosquitoes in New York State from 1993 to 2012. Oliver, J., G. Lukacik, J. Kokas, S. R. Campbell, L. D. Kramer, J. A. Sherwood, J. J. Howard. 2018. *Parasites & Vectors*, 11:362.

Raul Chuquiyauri

Affiliation: Medical Care Development International.

Education: MD from Universidad Peruana Cayetano Heredia, Peru. PhD from University of California at San Diego, CA, US. MPH from San Diego State University.

Research and Professional Experience: 15 years' experience in vector borne disease in the Peruvian Amazon. The last 10 years working in clinical trials, the last 2 years as Antimalarial Vaccine Clinical Director.

Professional Appointment: Clinical Director at Bioko Island Malaria Elimination Project, Equatorial Guinea

Publications from the Last 3 Years:

Antimalarial activity of single-dose DSM265, a novel plasmodium dihydroorotate dehydrogenase inhibitor, in patients with uncomplicated Plasmodium falciparum or Plasmodium vivax malaria infection: a proof-of-concept, open-label, phase 2a study. Llanos-Cuentas A, Casapia M, Chuquiyauri R, Hinojosa JC, Kerr N, Rosario M, Toovey S, Arch RH, Phillips MA, Rozenberg FD, Bath J, Ng CL, Cowell AN, Winzeler EA, Fidock DA, Baker M, Möhrle JJ, Hooft van Huijsduijnen R, Gobeau N, Araeipour N, Andenmatten N, Rückle T, Duparc S. *Lancet Infect Dis*. 2018 Aug;18(8):874-883.

Micro-heterogeneity of malaria transmission in the Peruvian Amazon: a baseline assessment underlying a population-based cohort study. Rosas-Aguirre A, Guzman-Guzman M, Gamboa D, Chuquiyauri R, Ramirez R, Manrique P, Carrasco-Escobar G, Puemape C, Llanos-Cuentas A, Vinetz JM. Malar J. 2017 Aug 4;16(1):312.

Single-Dose Tafenoquine to Prevent Relapse of Plasmodium vivax Malaria. Lacerda MVG, Llanos-Cuentas A, Krudsood S, Lon C, Saunders DL, Mohammed R, Yilma D, Batista Pereira D, Espino FEJ, Mia RZ, Chuquiyauri R, Val F, Casapía M, Monteiro WM, Brito MAM, Costa MRF, Buathong N, Noedl H, Diro E, Getie S, Wubie KM, Abdissa A, Zeynudin A, Abebe C, Tada MS, Brand F, Beck HP, Angus B, Duparc S, Kleim JP, Kellam LM, Rousell VM, Jones SW, Hardaker E, Mohamed K, Clover DD, Fletcher K, Breton JJ, Ugwuegbulam CO, Green JA, Koh GCKW. *N Engl J Med.* 2019 Jan 17;380(3):215-228.

Tafenoquine versus Primaquine to Prevent Relapse of Plasmodium vivax Malaria. Llanos-Cuentas A, Lacerda MVG, Hien TT, Vélez ID, Namaik-Larp C, Chu CS, Villegas MF, Val F, Monteiro WM, Brito MAM, Costa MRF, Chuquiyauri R, Casapía M, Nguyen CH, Aruachan S, Papwijitsil R, Nosten FH, Bancone G, Angus B, Duparc S, Craig G, Rousell VM, Jones SW, Hardaker E, Clover DD, Kendall L, Mohamed K, Koh GCKW, Wilches VM, Breton JJ, Green JA. *N Engl J Med.* 2019 Jan 17;380(3):229-241.

In: *Aedes aegypti*
Editor: Valèrie Megens

ISBN: 978-1-53618-197-5

Chapter 2

REVIEW ON *AEDES AEGYPTI*: ARBOVIRAL DISEASE TRANSMISSION AND CONTROL

***Edith Nonye Nwankwo**, PhD**
Department of Parasitology and Entomology,
Nnamdi Azikiwe University, Awka, Anambra State Nigeria

ABSTRACT

Aedes aegypti Linnaeus (Diptera: Culicidae) is a tropical mosquito and one of the most efficient mosquito vectors for the native African arboviruses. It is believed to have spread to other continents through trade and other human aid dispersal and now is distributed worldwide. It is a known nuisance species in the United States for centuries and is believed to have been brought to the new world on ships used for European exploration and colonization. Traditionally, *Ae. aegypti* was found in forested areas using tree holes as habitats. As a domestic breeder, it was initially found in breeding places on sailing ships from where it was distributed to all parts of the world. Adult *Aedes* mosquitoes show specific selection for sources of water or container types in which they lay their eggs. Hence, the species has a unique and specific environmental requirement for maintenance of its life cycle. The mosquito bites

* Corresponding Author's E-mail: chikanny@yahoo.com.

primarily during the day and is most active for almost two hours after sunrise and several hours before sunset, but it can bite at night in well lit areas. *Aedes* mosquito is known to bite people without being noticed because it approaches from behind and bites on the ankles and elbows, thus is referred to as an aggressive biter. It prefers biting people but it also bites dogs and other domestic animals, mostly mammals. It therefore represents a major threat to public health and inflicts terrible and unacceptable public health burden to humankind. This is due to its ability to transmit numerous human and animal pathogens such as Yellow fever, Dengue, Chikungunya and Zika viruses. *Aedes* mosquito control operations range from the use chemical control (larviciding or adulticiding), botanical control, bio-control, source reduction, mosquito traps among others. These control strategies are accomplished using pesticides, plant extracts and oil, biological-control agents, habitat modification and trapping. These are targeted to reduce nuisance mosquito cause to people around homes or in parks, recreational areas and their impact on real estate values, tourism, livestock and public health.

Keywords: *Aedes aegypti*, mosquito, vector, Arboviral diseases, vector control

Introduction

The largest tribe of mosquitoes is the Aedini, a clade of mosquitoes over 100 million years old with 1255 species in 10 genera (Jansen and Beebe, 2010). The most notable of which is the genus *Aedes*, presently recognized as containing 929 species (Akiner et al., 2016). *Aedes aegypti* (Diptera: Culicidae) is an important vector of arboviruses which originated in Africa (Central Africa) where ancestral populations are still present. The mosquito spread to other continents through trade and water barrels in ships and now is distributed worldwide (ECDC, 2012). This Invasive species have become established on six continents outside native ranges due to human-aided dispersal (ECDC, 2012). Historically, *Ae. aegypti* was found in forested areas breeding in tree holes but as an adaptation to urban domestic habitats, it is presently using a wide range of artificial containers such as vases, water tanks and tires (Jansen et al., 2010). The mosquito is also found utilizing underground aquatic habitats such as septic tanks and adapting to use both indoor and outdoor aquatic container habitats in the same area.

Adults of *Ae. aegypti* are relatively small and show a black and white pattern due to the presence of white/silver scale patches on a black background on the legs and other parts of the body. However, *it* could be mistaken for other invasive species such as *Aedes japonicus* and *Aedes cretinus* restricted to Cyprus, Greece, Turkey and *Ae. albopictus,* which was first noticed in Nigeria in 1991 after the outbreak of sylvatic (jungle) fever in many rural communities in Delta state in which the species was incriminated in the viral transmission of Sylvatic fever (Savage et al., 1992). The prevailing diagnostic character for *Ae. aegypti* is the presence of silver scales in a shape of a lyre on a black background on the dorsal part of the thorax (ECDC, 2012).

Adult *Aedes* mosquitoes characteristically hold their bodies low and parallel to the ground with the proboscis angled downward when landed. Females are distinguished by the shape of the abdomen, which usually comes to a point at its tip and by their maxillary palps (sensory structures associated with the mouthparts), which are shorter than the proboscis. They are larger than males and have small palps tipped with silver or white scales. Males have plumose antennae whereas females have sparse short hairs. Microscopic examination shows that male mouthparts are modified for nectar feeding and female mouthparts are modified for blood feeding. The proboscis of both sexes is dark and clypeus (segment above the proboscis) has two clusters of white scales.

Ae. aegypti is an aggressive daytime bitter and female bites primarily during the day and it is most active for approximately two hours after sunrise and several hours before sunset, but it can bite at night in well-lit areas (Das et al., 2018). It prefers mammalian hosts and preferentially feed on humans even in the presence of alternative hosts. The mosquito also feed multiple times during one gonotrophic cycle (blood feeding, egg-producing cycle), which has implications for disease transmission (Scott and Takken, 2012). The female mosquitoes need blood in order to produce eggs while male feed on nectar from plants. *Ae. aegypti*, females utilize many larval habitats such as temporary ground pools, temporary salt pools as well as rock pools for egg laying. Since the mosquito is common in areas lacking piped water systems, they depend greatly on water storage containers to lay their eggs (Aik et al., 2019). The females also lay their eggs singly on damp soils, tree holes that hold water intermittently as well as tide water pools in salt marshes, outflow from sewages, irrigated farmlands and rain water that have collected in ponds.

Aedes mosquito has dormant eggs, which can survive through long periods of drought up to nine months, allowing them to be easily spread to new locations. The eggs are reactivated by their immersion in water making the control of this species very difficult. With availability of required conditions, the eggs can hatch into larvae in less than a day or so and this could continue for several weeks and in extreme conditions, several months and even years. Egg hatching and survivorship depend on ambient temperature and relative humidity and can occur rapidly when level of water rises. Clemons et al. (2018) noted that mosquitoes in the tribe Aedini living in temperate areas typically overwinter in a diapaused egg stage that is resistant to freezing, which lasts until spring when they hatch as larvae.

Larvae of *Aedes* mosquitoes often wriggle to the water surface at intervals to obtain oxygen through their siphon, which serve as breathing tubes aiding them in breathing. Hence, in doing this, they lie vertically to the fresh water surface. The adult life span can range from two weeks to a month depending on environmental conditions. *Ae. aegypti* adult appears in three polytypic forms namely; domestic, sylvan, and peri-domestic.

The domestic form breeds in urban habitat, often around or inside houses. The sylvan form is a more rural form, and breeds in tree holes, generally in forests and the peri-domestic form thrives in environmentally modified areas such as coconut groves and farms (Kotsakiozi et al., 2019). *Ae. aegypti* is a known vector of the following four viruses that have been the most widespread and notorious in terms of severity of diseases and number of humans affected.

They include the viruses causing yellow fever (YFV), dengue fever (DENV), Chikungunya fever (CHIKV) and Zika (ZIKV) (WHO, 2014,

Souza-Neto et al., 2019). These arboviruses have been afflicting humans for millennia and have continued to cause immeasurable suffering. One mystery concerning these four viruses is that a single mosquito, *Aedes aegypti* out of more than 3,500 species has been the vector causing almost all major epidemics outside Africa (Scolari et al., 2019).

These arboviral diseases inflict a huge burden of morbidity and mortality worldwide and the principal control method is vector control (Wilson et al., 2020). Vector control is an effective way of reducing any vector-borne disease transmission and this is the primary objective of World Health Organization (WHO). Insecticidal measures are the most important means of controlling *Ae. aegypti* mosquito. Many insecticides of the group organochlorine (DDT), organo-phosphosphates (fenthion, Malathion and temephos), carbamates (bendicarb) and pyrethroides (permethrin deltamethrin, lambda-cyhalothrin) have been used for *Ae. aegypti* control (Nwankwo et al., 2019). In recent times, interventions using insecticides have been scaled up in many countries and the need to develop effective systems for insecticide management has been emphasized to ensure judicious use of insecticides. However, mosquito control has become increasingly difficult because of indiscriminate use of synthetic chemical insecticides which have adverse impact on the environment and disturb ecological balance. Majority of the insecticides are harmful to man and animals, some of which are not easily bio-degradable and spreading toxic effects. As result, several botanicals offer great promise as sources of phytochemicals for the control of *Aedes* mosquitoes (Rodriguez-Cavillo et al., 2019).

Six plant families with several representative species *Asteracae, Cladophoraceae, Labiatae, Meliaceae, ocystaceae* and *Rutacea*e, appear to have the greatest potential for providing future mosquito control agents. Some of these new compounds with their novel modes of bioactivity may prove useful in the development of safer insecticides for the future. Other environmentally-friendly control options for *Aedes* mosquito include; biocontrol, genetic control, use of mosquito traps, source reduction (habitat modification) among others (Wilson et al., 2020).

Global Distribution of *Aedes aegypti*

Traditionally, *Ae. aegypti* mosquitoes are known to have existing history in the tropics and subtropics. It is an African species which have invaded the New World through slavery, international trade and transport ships that resupplied in African ports (Powell and Tabachnick, 2013). Thus, making ports usually the first areas to be colonized by this mosquito. It has spread to all regions of the world, becoming the most frequently distributed species amongst other species in the genus (Brown et al., 2014). It is believed that *Ae. aegypti* invaded the rest of the world through this means and this method of spread presents significant tendency of introducing it into mainland of Europe from Madeira, the Netherlands and the North-Eastern Black Sea coast (Soumahoro et al., 2010). Since then, it has been reported to have re-invaded Madeira and some parts of Southern Russia and Georgia specifically Krasnodar Krai and Abkhazia (ECDC, 2016, ECDC, 2018) and in Netherland at Tire terminals, making it amongst the most widespread mosquito species globally.

There are no existing facts based on climate on why *Ae. aegypti* when introduced in Europe could not thrive towards Southern Europe but if this is enacted could bring an increase on the risk of transmission of commonly known arboviruses.

Although it is limited due to its sensitivity to temperate winters but over the past 25 years, there has been an extensive global spread (ECDC, 2016). *Ae. aegypti* exhibits ecological plasticity and is currently distributed in Africa, the surrounding tropics and subtropics, south eastern USA, the middle east, south east Asia, Indian island and northern Australia (Soumahoro et al., 2010, ECDC, 2016). It is reported to have been introduced into Netherlands and Africa (Weetman et al., 2018) with shipped tyres from Florida, USA (Brown et al., 2011). The mosquito has currently established in all parts of the world with the exception of Antarctica. The wide spread of this mosquito has allowed the entry and epidemics of arbo-viruses such as yellow fever, dengue, chikungunya and zika virus into new populations of susceptible human hosts (WHO, 2018).

Biology of *Aedes aegypti*

Aedes aegypti is a holometabolous insect that goes through a complete metamorphosis (egg, larva, pupa, and adult stage) having a complex life-cycle with dramatic changes in shape, function and habitat. After taking a complete blood meal, females produce on average 100 to 200 eggs per batch; but the number of eggs produced is dependent on the size of the blood meal, thus, a smaller blood meal produces fewer eggs (Nelson, 1986). Females can produce up to five batches of eggs during a lifetime.

Females lay their eggs on the inner, damp or wet walls of containers with water in areas that are temporarily flooded, such as tree holes and man-made containers (Bowls, cups, fountains, tires, barrels, vases and any other container storing water), and are laid singly rather than in a mass (CDC, 2012). It takes only a very small amount of water to attract a female mosquito. The eggs are hardy and stick to the walls of a container like glue and can survive drying out for up to 8 months even over the winter in the southern United States. Most often, eggs are placed at varying distances above the water line, and a female will not lay the entire clutch at a single site, but rather, spread out the eggs in two or more sites over hours or days, depending on the availability of suitable substrates (Foster and Walker, 2002). Eggs of *Ae. aegypti* are long, smooth, ovoid shaped and approximately one millimeter long. When first laid, eggs appear white but within minutes, turn shiny black. In warm climates such as the tropics, eggs may develop in as little as two days whereas in cooler temperate climates, development can take up to a week (Foster and Walker, 2002). The eggs can survive desiccation for months and hatch once submerged in water making the control of *Ae. aegypti* very difficult. Larval development is also temperature dependent and larvae are often found around the home in puddles, tyres or within any object holding water. The larvae pass through four instars, spending a short amount of time in the first three and up to three days in the fourth instar.

Most *Aedes* larvae can be distinguished from other genera by unaided eye using their short posteriorly located siphon. Larvae breathe oxygen through the siphon, which is held above the water surface while the rest of the body hangs vertically.

If temperatures are cool, *Ae. aegypti* can remain in the larval stage for months so long as the water supply is sufficient (Foster and Walker, 2002, Lippi et al., 2019). The development of the larval stages lasts for at least four days and is spent at the water surface as larvae need to breathe; although they wriggle sporadically to the bottom of the container when disturbed and as such are called "wrigglers" or "wigglers,." Larvae also swim below the surface when they feed on organic solid or liquid matter such as strands of macro-algae, leaves, dead invertebrates which can be of their own kind and other microscopic organisms in their aquatic habitats.

Following the fourth larvae instar, *Aedes* species enter a mobile pupae and respond to stimuli (characteristics that distinguish it from other holometabolous insects). Pupae, also called "tumblers," is a non-feeding stage which takes approximately two days to develop (Clemons et al., 2018). The pupae acquire oxygen through organs known as respiratory trumpets and when disturbed, they exhibit a jerky fashion movement up and down in the water. Hence, the pupae is adapted to progression in an upward zigzag by alternate flexion and relaxation of the abdomen. On emergence, adults take in air to expand the abdomen and split open the pupal case along the mid-dorsal line of the cephalothorax and emerge head first. Males develop faster than females, so males generally pupate earlier (Steinwascher, 2018). Adult that emerges newly stays on the surface of water often for a short period to allow its cuticle to dry and harden and air pressure in the stomach to expand the wings and legs. After few minutes, the adult is able to leave the water surface and fly off only for a close distance.

Most adult mosquitoes fly only for a little distance such as a few hundred metres or less from sites where they emerged (Service, 1996). Although the flight habits of mosquitoes vary amongst species, most species found around human habitations live closely to their emergence point while others fly far away from their breeding sites. For instance, *Aedes* mosquitoes migrate many kilometres away from breeding sites although the flight limit for females is usually longer compared to those of males. Certain factors help the flight and migration of mosquitoes, one of which is wind, considering that most mosquitoes stay within 1.61 km while others recorded 120.75 km away from place of breeding (Provonsha, 1983).

Three days after emergence, male mosquitoes feed on nectar from flowers while adult female feed on humans and animals for blood in order to produce eggs. Some female mosquitoes can feed on only one type of animal or prefer to feed on various animals, thus, capable of disease transmission. Most female mosquitoes have to obtain enough or adequate blood from man or other animals before the eggs can develop. Where they are not able to achieve this, they die having laid no viable eggs. However, some species of mosquitoes have also surprisingly evolved means of laying viable eggs without even obtaining blood meal. Subsequently, the female lays eggs to continue the cycle again (CDC, 2012). The average life span of an adult *Aedes* mosquito in nature is 2 weeks during which a female is capable of laying eggs thrice in its life time producing about 100 eggs each time. Mosquito incubation process largely depend on temperature and is generally restricted to 14-30°C while in the tropical regions, advancement from egg to adult can take 7 to 13 days. According to CDC (2012), the entire life cycle lasts 8-10 days at room temperature depending on the level of feeding. Thus, there is an aquatic phase (larvae, pupae) and a terrestrial phase (eggs, adults) in the *Ae. aegypti* life-cycle.

Breeding Habitat of *Aedes aegypti*

Mosquito breeding habitats can be natural or man-made. *Ae. aegypti* breeds and develops in artificial containers of small volume such as flasks, bottles, flower vases, tin cans, jars, discarded automobiles tyres, unused water closets, cisterns, rain barrels, sagging roof gutters and in natural sites such as coconut shells, snail shells, leaf axils and tree holes. They are generally container inhabiting mosquitoes and thrive in densely populated areas in close contact with people making them an exceptionally successful vectors of arbo-viruses. They are extremely successful in areas which lack reliable water supplies, waste management, sanitation and they depend greatly on stored water for breeding sites (ECDC, 2016). *Aedes* mosquito proliferates well in clustered areas or slums with unreliable, inadequate or insufficient supplies of water, poor waste management and environmental sanitation.

A study by Huang (2004) observed that discarded tins were the most prolific source of *Ae. aegypti* larvae compared with other breeding habitats such as oil drums, leaf axils, old tyres and water storage containers. Similarly, Trpis (1972) identified tyres, tins, wrecked motor cars, water-pots, snail shells, coconut shells, and tree holes as the most common breeding sites of *Ae. aegypti* of which tyres were by far the most important and provided a constant source of *Ae. aegypti*. It was also found that *Ae. aegypti* mosquitoes were abundant in urban areas and this was linked to availability of breeding habitats in urban environment. Since the mosquitoes mostly thrive in urban areas in close contact with people, humans are exceptionally successful hosts of this mosquito.

Aedes mosquitoes are also found in ground water (temporary and specialized but not permanent) habitats requiring standing water for the completion of their growth cycle with their potential breeding sites being places that contain standing bodies of water (Ferede et al., 2018).

The adult female *Aedes* mosquitoes are either domestic, breeding around houses or in the jungle (wild) or in both habitats (semi-domestic). The egg stage of mosquitoes in their breeding site is very important because its site can be used to determine the characteristics of that particular species. Also, distribution of eggs would increase the probability that some of the offspring would survive. Therefore, blood-fed female mosquitoes can have their eggs distributed over several sites with an average of 11 eggs or more per site. The presence or absence of conspecific eggs influences the number of eggs laid at a given site.

Behaviour of Adult *Aedes* Mosquitoes

Ae. aegypti mosquito is generally considered an urban mosquito, being most frequently found in urban and sub-urban areas in close proximity to humans (highly anthropophilic) and where number of houses are high, preferring to live indoors (endophilic) and being frequently caught inside houses feeding on human blood (endophagic). *Aedes* mosquitoes are painful, aggressive, persistent and diurnal biters with biting peak at dawn and dusk. They occasionally enter dwellings to bite with preference for mammals including man (Provonsha, 1983).

Also, *Ae. aegypti* affinity for close association with man and living around human habitations in West Africa, and its extensive dispersion and settlement in the tropical regions led to the significant effective inter-human transmission of arboviruses (Weaver and Reisen, 2010).

In this region, only some species such as *Diceromyia* (*furcifer-taylori* group) and *Stegomyia* (*africanus, luteocephalus* and to a lower extent *aegypti, vittatus* and *metallicus*) prefer man and probably also monkeys to other vertebrates. Majority of other known species obtain blood meals from domestic and wild animals and bite man by chance or accidentally. Studies have shown that females of 46 *Aedes* (*Stegomyia*) species and subspecies (such as *aegypti, formosus, africanus, luteocephalus, woodi, granti, albopictus* and *unilineatus*) are known to feed on humans (Huang, 2004). Contrary to this, Vainio and Cutts (1998) reported that the *Ae. aegypti aegypti* (domestic form) is anthropophilic while *Ae. aegypti formosus* (wild form) is zoophilic subspecies of *Aedes aegypti*.

Hamon (1973) is of the opinion that *Aedes* mosquito is man-biting species in and around villages and in open areas, they bite during the day with their peak of activity at dusk. This is when humans usually relax out of doors. This ensures a high degree of contact between mosquitoes and one of their preferred hosts, as almost all African *Aedes* species are exophageous. On the contrary, the zoophilic species present a more even level of activity all the night long or have a peak of activity in the middle of the night. This was supported by Huang et al. (2010), who noted that the adult females of *Ae. fryeri* mostly bite at night, although they can also bite at any time of the day, with two clear biting peaks occurring at sunset and dawn. The evening biting peak occur between 18:00 and 19:00 hr. of the day and is always higher than the morning peak, which slightly had longer duration between 05:00 to 07:00 hr. This mosquito can bite people without being noticed because it approaches from behind and bites on the ankles and elbows. During passive times, adult females of *Ae. aegypti* typically rests indoors in dark corners of bedroom especially behind clothes or other concealed areas (WHO, 2009).

It was also found that most of them rest on temporary objects (clothing and mosquito nets), while a few percentage was found resting on furniture and other permanent articles.

ARBOVIRAL DISEASES TRANSMITTED BY *AEDES AEGYPTI* MOSQUITO

Diseases caused by *Ae. aegypti* are increasing global public health concerns owing to their rapid geographical spread and increasing disease burden. This is as a result of ongoing range expansion fueled by increased global trade and travel. The mosquitoes are significant vectors of various human diseases, as they can transmit extremely large number of arboviruses that are responsible for substantial human morbidity, suffering and mortality nearly worldwide. *Ae. aegypti* is considered to be a competent vector experimentally of at least 22 arboviruses including Japanese encephalitis virus, sindbis virus, potosi virus, cache valley virus, La crosse virus, Eastern equine encephalitis virus among others (Eritja, 2005). According to Weaver and Reisen (2010), Powell (2018), most of the *Aedes* transmitted viruses are flaviviruses namely; dengue (DEN), Yellow fever (YF) and Zika; alphaviruses namely Chikungunya (CHIK) and Venezuelan equine encephalitis (VEE) (Mayer et al., 2017). The CHIK virus was historically found only in the Old world while VEE virus was found only in the new world. DENV virus has a nearly widespread or very prevalent distribution in the tropics and more recently, it has been introduced into Europe, becoming the most prevalent arboviral infection affecting man, causing 100 million yearly infections globally with up to half of the world's population predisposed to possible infection (Brady et al., 2012). The blood feeding habit of female *Aedes* species supports easy disease transmission of these arbo-viruses.

Ae. aegypti is considered one of the potential vector for arboviruses because of its high affinity for human blood, frequently biting various times before completion of egg development, and always lives in close association with man. Biological transmission of virus can be from carrier or infected female to its offspring (vertical transmission) or from a carrier or infected male linearly to a female counterpart (horizontal transmission). Oral transmission can also emanate from an infected female vector to animals or humans through injecting saliva while blood-feeding (horizontal transmission) (Weaver and Reisen, 2010). The oral mode of transmission is the commonest form of transmission for most arboviruses. This involvs vector infection through blood meal, amplification of the virus in the salivary system and introduction of virus into various hosts during intermittent blood meals (Weaver and Reisen, 2010). In addition to horizontal transmission of the disease, vertical transmission during pregnancy or at birth has been reported in man, while transovarian transmission has been reported in mosquitoes. This further facilitates the spread of the disease and increase the difficulties in combating them. These arbo-viral diseases are as follow.

Yellow Fever Virus

Yellow fever (YF) is often a classic haemorrhagic fever whose history traces back to the history of the New World. It is an acute infectious disease and one of the most lethal viral diseases caused by a mosquito-borne RNA virus of the genus Flavivirus from the family Flaviviridae (NCDC, 2019). The mosquito vector carries the virus that causes the disease from one host to another, primarily between monkeys in the rainforest zone, from monkeys to humans during the intermediate cycle in moist savanna or forest savanna ecotone, and from person to person in the dry savanna and urban areas (Nwankwo et al., 2019).

In rural human settlements, YF is transmitted through the bites of infected female *Aedes* mosquitoes which include *Aedes aegypti*, *Ae. africanus*, *Ae. opok*, *Ae. simpsoni s.l.*, *Ae. luteocphalus*, *Ae. tylori*, *Ae metallicus*, *Ae. furcifer*, *Ae. vittatus*, and by *Ae. aegypti* in urban settings (Barret and Higgs, 2007). The risk of this disease transmission could be an accumulation of abundant presence of mosquitoes, biting preferences and biting patterns (Bustamante and Lord, 2010).

Yellow fever epidemics still occur frequently in the tropics, and can occur in temperate regions during summer months, although it is not the major threat it once was (Timori, 2004). In some patients, a toxic phase follows in which liver damage with jaundice can occur and eventually lead to death. WHO estimates that yellow fever causes 200,000 illnesses and 30,000 deaths every year in unvaccinated population, of which around 90% of infection occur in Africa (Vainio and Cutts, 1998, NCDC, 2019). Today, there is a very successful vaccine for yellow fever, which has contributed to the decline in cases. In 1951, Max Theiler won a Nobel Prize for his vaccine, which is the only Nobel Prize given for a vaccine to date (Norrby, 2007).

There have been achievements in understanding the spread and control of yellow fever (YF) disease which has also led to the manufacture of a readily obtainable safe and effective vaccine.

However, this disease remained a significant and important public health issue in both Africa and the Americas where about 200,000 persons were affected annually with 30,000 deaths occurring worldwide (WHO, 2013). Forty-seven countries in Africa and Central and South America were either endemic for, or have regions with high prevalence of yellow fever (WHO, 2017).

In 1983 - 1984, there were outbreaks in several districts of Burkina Faso (Fada N'Gourma, Manga, Ouagadougou and Tenkodogo) which spilled into the Northern and Upper West regions (Jirapa and Wa districts) of Ghana, spreading to the Eastern parts of the Upper East region. It affected 19 villages with 28 cases and 5 deaths (WHO, 2005, 2012). Also, a study based on African data sources assessed the burden of yellow fever during 2013, reporting about 84, 000 - 170, 000 severe cases of the disease and 29, 000 - 60, 000 deaths caused by the yellow fever virus (WHO, 2017). The Nigeria Centre for Disease Control (NCDC) in 2017 officially notified a confirmed case of yellow fever in Kwara state to WHO as per the International Health Regulations (2005). The country has since then been responding to successive yellow fever outbreaks over a wide geographic area. In 2019, Nigeria reported an outbreak of yellow fever with an epi-centre in the Yankari game reserve of Alkaleri LGA, Bauchi state. According to Nigeria Centre for Disease Control (NCDC, 2019), 231 suspected cases have been reported in four states including Bauchi (110), Borno (109), Gombe (10), and Kano (2).

Although yellow fever disease has never been reported in Asia, the continent is at risk because the conditions required for possible transmission, such as increase in human population and movement of the *Aedes* vectors into urban and peri-domestic environments, are promisingly present there (Weaver and Reisen, 2010).

Irregular visits by travelers to countries where yellow fever is highly prevalent have the possibility of bringing the disease to countries free from or with no case of yellow fever. In order therefore to interrupt such importation of the disease or the virus, several nations now request for evidence of vaccination against the virus before they endorse or approve a visa, especially a situation where travelers come from, or have visited yellow fever endemic areas.

Dengue Fever Virus

Dengue fever is amongst the most important and fastest vector-borne arboviral infection transmitted to humans. It is a single stranded RNA virus belonging to the family Flaviviridae (Gubler, 1997, Vidyadhara and Naveen, 2018). During the 19th century, dengue was considered a benign sporadic disease that caused epidemics at long intervals. However, in the past five decades, the incidence was reported to have increased 30-folds.

An estimated 100 million cases of dengue fever, 500,000 cases of dengue hemorrhagic fever and 25,000 deaths have been reported annually (Gubler et al., 1998, WHO, 2012a). According to WHO (2000), dengue fever and dengue haemorrhagic fever (DHF)/dengue shock syndrome (DSS) occur in 124 countries where they are threat to the health of more than 55% of the world's population living in urban, suburbs and rural areas of the tropical and subtropical regions of the world. Recent report by WHO (2019) shows that the global incidence of dengue has grown dramatically and there is an estimated 390 million infections each year.

In humans, dengue infection ranged from simple fever to much more severe and sometimes fatal dengue haemorrhagic fever (DHF) and dengue shock syndrome (DSS) (WHO, 1995). It is largely prevalent in places like Africa, the Americas, Eastern Mediterranean, South-East Asia, and the Western Pacific, threatening more than 2.5 billion people with 1.2 million cases in 2008 and over 3.34 million in 2016 (WHO, 2019).

The significant burden of the disease is felt in South-East Asia and the Western Pacific (WHO, 2009). In Southeast Asia, dengue haemorrhagic fever (DHF) is a major cause of mortality among children while dengue fever (DF) occurs epidemically affecting older children and adults. The outbreaks are often described as explosive.

Attributive and associated risk factors to the increased cases of dengue include increase in the growth of human population, increased migration, finite financial and human resources. These factors led to emergence and subsequent occurrence of the disease (Gubler, 2004). DEN has a nearly widespread occurrence in the tropical regions of the world and has lately been reported in Europe (ECDC, 2012b; Schaffner and Mathis, 2014).

DEN viruses that cause majority of diseases to humans are not zoonotic but entirely exploit humans as reservoir and amplification hosts (Weaver and Reisen, 2010). DEN viruses however depend heavily on mosquitoes which live in close contact or affiliation with people for possible transmission of which *Ae.* (*Stegomyia.*) *aegypti* is the primary vector in most places (almost all countries) with *Ae.* (*Stg.*) *albopictus* playing a role as an auxiliary vector in a number of places. *Ae. africanus* and *Ae. luteocephalus* also act as potential vectors in Africa. It is also possible that infected females of *Ae. aegypti* and *Ae. albopictus* can transmit the virus to their next generation transovarially (da Costa, 2017). Once a mosquito contracts the virus, it remains infected throughout its lifespan and can inject or pass on the virus while obtaining blood meals (da Costa, 2017). Dengue fever virus is commonly maintained or preserved in an *Ae. aegypti-Human-Ae. aegypti* cycle with known repeated epidemics.

While the United States rarely experiences yellow fever cases, the most recent concern to the United States and Florida is the transmission of dengue virus. Dengue is also known as "break-bone fever" for the excruciating pain victims feel. It is a dangerous disease due to four different serotypes: DEN-1, DEN-2, DEN-3, and DEN-4 (WHO, 2019). Although a person can obtain immunity to one serotype, they are still susceptible to the others. The most deadly is dengue hemorrhagic fever (DHF), which is often fatal. In 1981, Cuba experienced a major DHF outbreak, killing 159 people (Bravo et al., 1987). Though outbreaks in the United States are still rare, Mexico is experiencing major dengue outbreaks, and its close proximity could lead to outbreaks in the United States. Like yellow fever, dengue fever is caused by a flavivirus (attacks the liver) and can only be transmitted by female mosquitoes.

Although there has not been any report of cases of DEN in Ghana, it has been detected and reported in La Côte d'Ivoire (2006 and 2008, DEN-2 and DEN-3) and Burkina Faso (1982, Den-2), which borders with Ghana (WHO, 1995). In Nigeria, evidence on the prevalence of DENV infections from more robust and generalised surveys is lacking. However, this disease was recently detected in Cross River state, south-eastern part of the country (Otu et al., 2019). Despite the endemicity of dengue in Nigeria, the disease is not widely reported and routine diagnosis is neglected. With increase in migration of people across countries belonging to ECOWAS and deficiency of mosquito control programmes in these areas, it is possible that the situation of viral haemorrhagic fever transmission in the region may be heightened, hence the need for vector surveillance.

Unfortunately, there is no effective vaccine against dengue virus, which would ideally provide protection against all four serotypes. As a consequence, controlling the spread of dengue requires that the mosquitoes be targeted directly. There are also periodic health campaigns to find solution for the problem. Nevertheless, these activities are carried out in accordance with policies of the countries affected by the problem. However, all attempts are focused on the population decline of the mosquitoes. Actions includes environmental sanitation, removal of breeding areas, application of chemical insecticides, use of long lasting insecticide treated nets. Also, application of biological control with fungi, nematodes or predators such as fish, lizards and gecko have been utilized. Another approach is the use of genetically modified mosquitoes, some of which are engineered to be genetically sterile and hence induce population suppression.

Zika Virus

Zika virus (ZIKV) belongs to the family of virus known as Flaviviridae and the genus Flavivirus (Malone et al., 2016). Its name originates from a forest in Uganda known as Zika Forest, where the virus was first reported in 1947 from a sentinel rhesus monkeys and subsequently from mosquito (Sikka et al., 2016, Wikan and Smith, 2017). Thereafter, first three cases of the virus was reported in humans five years later in Oyo state, Nigeria (MacNamara, 1954). Fagbami (1979) reported in his study that 40% of Nigerian adults and 25% of Nigerian children have antibodies to Zika virus. Since the 1950's, the infection has been present inside a confined tropical belt from Africa to Asia.

Between 2007 and 2016, spread of the virus has been moving eastwards towards the Pacific Ocean and up to the Americas, where the 2015 - 2016 Zika virus widespread occurrence was reported to have reached pandemic levels (Smith, 2016). Prior to 2007, cases of ZIKV human infections have been rare and sparse and without pattern in tropical Africa and a few places in Southeast Asia. But since 2007, various outbreaks have been reported towards the Pacific Islands showing the viral spread outside its previously documented geographic coverage. Native or indigenous transmission of ZIKV in South America was reported in early

2015 (Campos et al., 2015) with widespread prevalence reported in 2015 in Central and South America and the Caribbean (Petersen et al., 2016).

Petersen et al. (2016) reported that thirty-two nations and regions of the Americas were hit by Zika, this also include Rio de Janeiro, Brazil, where the Olympics and Paralympics games scheduled for August 2016 were held; this definitely posed imminent danger to foreign travelers, local organizers and residents. Since 2015, more than 70 countries and territories have reported local Zika virus transmission (WHO, 2016) and the most of the reported cases of Zika virus are known to exist in the tropics such as Brazil, Colombia, Paraguay, Suriname, Venezuela and French Guiana. Areas found within the USA which are of concern for potential Zika-infected mosquitos are those of wet lowlands, warmer temperatures and higher levels of poverty (CDC, 2017). Recent report in the Republic of Congo showed that the majority of the population are immunologically naïve against ZIKV with a presumptive (peri-) sylvatic transmission cycle (Nurtop et al., 2020).

The virus is commonly transmitted by infected adult female *Ae. aegypti* mosquito although report exists of the presence of the virus in common *Culex* house mosquitoes (*Culex quinquefasciatus*) (Franca et al., 2016). The virus has also been detected in a few tree-hole living mosquito species in the *Aedes* genus, such as *Ae. africanus*, *Ae. apicoargenteus*, *Ae. furcifer*, *Ae. hensilli*, *Ae. luteocephalus* and *Ae. vittatus*, with a development period of about 10 days in mosquitoes (Hayes, 2009).

In Africa, Zika has been reported and isolated in mosquitoes of the genera *Aedes*, *Anopheles* and *Mansonia* in Senegal, La Côte d'Ivoire, Burkina Faso, Gabon and the Central African Republic (Vorou, 2016) with no case reported in Nigeria yet. However, the wide distribution of the virus in animal hosts and vectors in Nigeria favours the rise of recombinants and creates concern for surveillance and vector control especially in the country.

Chikungunya Virus

The term Chikungunya often refers to both the virus (CHIKV) and the illness or fever (CHIKF) caused by this virus. It was derived from the African dialect Swahili or Makonde and translates as "to be bent over." Chikungunya fever is a self-remitting febrile viral illness caused by RNA alphavirus that has been associated with frequent outbreaks in tropical countries of Africa and Southeast Asia (Natesan, 2019, Rezza et al., 2019). The illness has only recently become a concern in Western countries and temperate zones around the world. The recent re-emergence and travel-related spread of Chikungunya infection to Europe and the United States has drawn global attention with international travel being prominent as one of the major risk factors for the rapid global spread of the disease.

CHIKV received significant public health interest due to the outbreaks in Réunion Island, southwestern part of Indian Ocean in 2005-2006 where 225,000 cases of infections was reported (Borgherini et al., 2007), Italy in 2007 affecting a total of 205 persons (Rezza et al., 2007) and France in 2010 and 2014 with 2 and 11 locally transmitted cases recorded respectively (Paty et al., 2014). Recent invasion into the Americas has also been reported with over 1 million cases recorded as of 2014.

Although chikungunya is not endemic in North America like dengue, the number of cases is steadily increasing and this virus could become a major threat to public health in the United States. Most cases documented in the United States are associated with international travel, but with the spread and resurgence of the yellow fever mosquito and Asian tiger mosquito in the Americas, chikungunya is a very real threat (CDC, 2017). CDC (2017) reported that a total of 156 chikungunya virus disease cases with illness onset in 2017 have been reported to ArboNET from 28 U.S. states and all reported cases occurred in travelers returning from affected areas. No locally-transmitted cases have been reported from U.S. states.

CONTROL OF *AEDES* MOSQUITO

Mosquito control is aimed at managing the population of mosquitoes to reduce their damage to human health, economies and enjoyment. Mosquito control is a vital public-health practice throughout the world especially in the tropics because mosquitoes spread many arboviral diseases. Vector control is a major component of the World Health Organization global control strategy and when implemented properly, it can be an effective strategy for preventing mosquito borne disease transmission (WHO, 2015).

Chemical insecticides are the mainstay in the fight against vector-borne diseases (WASH, 2016). Larval breeding sites should be searched and treated with insecticides as frequent as possible by trained field technicians and trained community members. Mosquito elimination in larval stages before emerging to adults will reduce the adult mosquito population. Control of larvae can also be accomplished through use of growth regulators, surface films, stomach poisons (including bacterial agents), and biological agents such as fungi, nematodes, copepods, and fish (Thomas, 2017).

A growth regulator commonly used is methoprene, considered slightly toxic to larger animals, which mimics and interferes with natural growth hormones in mosquito larvae, preventing development. Reduction of mosquito breeding sites such as jars, barrels, pots, vases, bottles, tins, water coolers, and tyres can be done by environmental management, removing of solid waste and managing artificial man-made habitats. Also, all domestic water storage containers should be cleaned and covered daily.

Control of adult mosquitoes is the most familiar aspect of mosquito control to most of the public. The use of insecticides such as lambda cyhalothrin- or deltamethrin-treated material by hanging them on windows and used as water jar covers may reduce adult *Ae. aegypti* population. Adult mosquito control is also accomplished by ground-based applications or through aerial application of residual chemical insecticides. In developed countries, modern mosquito-control programs use low-volume applications of insecticides, although some programs may still use thermal fogging. Besides fogging, there are some other insect repellent for indoors and outdoors.

Application of repellents such as DEET, DIMP are important in reducing human to vector contact and these are done during active hours of the day (Youdeowei and Service, 1995). Generally, *Ae. aegypti* is a highly resilient and difficult to control because of its adaptable nature and efforts to eradicate it have failed. However, this species are managed through the implementation of the following control tactics:

Chemical Control

Chemical control remains the most widely used approach for vector control (WHOPES, 2011). Insecticidal measures are the most important means of control of *Ae. aegypti*, the main vector of arboviruses especially in the outbreak-risk areas. Many insecticides of the group organochlorine (DDT), organo-phosphorous (fenthion, Malathion and temephos), carbamates (Bendiocarb) and pyrethroides (permethrin deltamethrin, Lambda-cyhalothrin, etc.) have been used in mosquito control programmes and for *Ae. aegypti* control (WHO, 2015). DDT was formerly used throughout the world for large area mosquito control, but it is now banned in most developed countries. However, DDT remains in common use in many developing countries (14 countries were reported to be using it in 2009, which claim that the public-health cost of switching to other control methods would exceed the harm caused by using DDT. It is sometimes approved for use only in specific, limited circumstances where it is most effective, such as application to walls.

The role of DDT in combating mosquitoes has been the subject of considerable controversy. Although DDT has been proven to affect biodiversity and cause egg shell thinning in birds such as the bald eagle, there is belief that DDT is the most effective weapon in combating mosquitoes. There is also some disagreement based on discrepancies in the extent to which disease control is valued as opposed to the value of biodiversity. This has genuinely be argued amongst experts as per the costs and benefits of using DDT.

Unfortunately, DDT-resistant mosquitoes have started to increase in numbers especially in the tropics due to mutations and indiscriminate application of pesticides (Nwankwo et al., 2019). In areas where DDT resistance is encountered, malathion, propoxur or lindane are used as alternatives.

Although synthetic organic insecticides are highly efficacious against target species such as mosquitoes (WHO, 2006), they can be detrimental to a variety of animal life including man. The continuous use of various kinds of insecticides has increased mosquitoes' resistance. This action causes low insecticidal susceptibility in the mosquitoes and thus contributes to further development of their population. In addition to adverse environmental effects from conventional insecticides, most major mosquito disease vectors have become physiologically resistant to many of these compounds. These factors have created the need for environmentally-safe, biodegradable and target-specific insecticides against mosquitoes and the search for such compounds has been directed extensively to the plant kingdom.

Botanical Control of *Aedes* Mosquito

The use of different parts of locally available plants and the various products in the control of mosquitoes has been well established globally by numerous researchers. The larvicidal properties of indigenous plants have also been documented in many parts of Nigeria (Nwankwo et al., 2011, Nwankwo et al., 2012; Okonkwo et al., 2014). Phytochemicals derived from various botanical sources have provided numerous beneficial uses ranging from pharmaceuticals to insecticides.

In recent times, natural products from plants are widely under investigation against insects due to their excellent properties, cheap availability, renewable nature, environmentally-friendly nature and the presence of an array of characteristics such as insecticidal, antifeedant, ovicidial and larvicidal effects (Nwankwo et al., 2015). Herbal products with proven potential as insecticide or repellent can play an important role in the interruption of the transmission of mosquito borne diseases at individual as well as at community level.

Plant parts and derivatives have been traditionally used by human communities to control these mosquitoes. Azadirachitin, anabasine, quassia, nicotine, pyrethrin, D-liminene and carvacrol were important active components of plants used before the advent of synthetic chemicals. Essential oils obtained from leaves of different plants have been reported to possess larvicidal activities against *Aedes aegypti* (Nwankwo et al., 2015). *Psidium guajava* plant of the family Myrtaceae has shown insecticidal activities. It is described as a low evergreen tree or shrub 6-25 feet high with wide spreading branches and downy twigs. Studies on the aqueous and ethanolic extracts of *P. guajava* has revealed that the extracts have larvicidal potential and have shown effectiveness in the control of *Ae. aegypti* (Rwang et al., 2016). *Terminalia catappa* of the family Combretaceae is widely grown in tropical regions of the world as an ornamental tree and the fruit is edible. Its extracts has been proven to be potent insecticides, feeding deterrents and progeny production inhibitors. Research has shown that *T. catappa* can be a potential candidate against *Ae. aegypti* particularly in its larvicidal and pupicidal activities. Ethanol extract of the dried leaf plant has shown maximum larvicidal activities (Rodriquiz-Cavallo et al., 2019).

Biocontrol

Biological control or "biocontrol" is the use of natural enemies to manage mosquito populations. There are several types of biological control including the direct introduction of parasites, pathogens and predators to target mosquitoes (Yan-Jang et al., 2017). Effective biocontrol agents include predatory fish that feed on mosquito larvae such as mosquito fish (*Gambusia affinis*) and some cyprinids (carps and minnows) and killifish (Kramer et al., 1988, Tietze et al., 1991).

Tilapia also consume mosquito larvae when utilized in a controlled system through aquaponics. This is known to provide mosquito control without adverse effects on the ecosystem. Other predators include dragonfly (fly) naiads, which consume mosquito larvae in the breeding waters, adult dragonflies and some species of lizard and gecko, which eat adult mosquitoes. Biocontrol agents that have had lesser degrees of success include the predator mosquito, Toxorhynchites (Yasuno and Tonn (1970) and predator crustaceans-Mesocyclops copepods. Predators such as birds, bats, lizards, and frogs have been used, but their effectiveness is only anecdotal. Microbial pathogens of mosquitoes include viruses, bacteria, fungi, protozoa, nematodes and microsporidia (Kramer et al., 1988). Dead spores of the soil bacterium, *Bacillus thuringiensis*, especially *Bt israelensis* (BTI) interfere with larval digestive systems. It can be dispersed by hand or dropped by helicopter in large areas. Two species of fungi that are known to kill adult mosquitoes are *Metarhizium anisopliae* and *Beauveria bassiana* (Giovannie et al., 2016, Evans et al., 2018).

Genetic Control

Introducing large numbers of sterile males is another approach to reducing mosquito numbers. Sterilization using chemical compounds such as thiotepa was successful but also led to the potential of environmental contamination. A modern adapted version of SIT has been successfully developed to target *Ae. aegypti* for the control of DENV, CHIKV, and ZIKV (Phuc et al., 2007). The technique does not rely on the early-acting lethality as SIT, but depends on the late-acting lethality to prevent the production of viable progenies. The approach is based on release of insects with dominant lethality (RIDL), which introduces the expression of a dominant lethal gene in the vector populations by releasing male transgenic mosquitoes carrying the dominant lethal gene.

Another control approach under investigation for *Aedes aegypti* uses a strain that is genetically modified with antibiotic tetracycline. Modified males develop normally in a nursery while they are supplied with this chemical and can be released into the wild (Morris et al., 1989). However, their subsequent offspring will lack tetracycline in the wild and never mature. Field trials were conducted in the Cayman Islands, Malaysia and Brazil to control the mosquitoes that cause dengue fever. In April 2014, Brazil's National Technical Commission for Biosecurity approved the commercial release of the modified *Ae. aegypti* mosquito (de Andrade et al., 2016). Other genetic methods tested in the same year include cytoplasmic incompatibility, chromosomal translocations, sex distortion and gene replacement (de Andrade et al., 2016). The researchers were of the view that several years away from the field trial stage, if successful, these other methods have the potential to be cheaper and to eradicate the *Ae. aegypti* mosquito more efficiently.

Mosquito Trap

A traditional approach to controlling mosquito populations is the use of ovitraps or lethal ovitraps, which provide artificial breeding spots for mosquitoes to lay their eggs. While ovitraps only trap eggs, lethal ovitraps usually contain a chemical inside the trap that is used to kill the adult mosquito and/or the larvae in the trap. Lethal ovitraps are considered a sustainable and environmentally friendly method of controlling container-inhabiting *Aedes* mosquitoes (Long et al., 2015). Studies have shown that with enough of these lethal ovitraps, *Aedes* mosquito populations can be controlled (Long et al., 2015). In Australia, dengue control combines source reduction with lethal ovitraps to reduce *Ae. aegypti* populations during outbreaks. A recent approach is the automatic lethal ovitrap, which works like a traditional ovitrap but automates all steps needed to provide the breeding spots and to destroy the developing larvae (Long et al., 2015).

In 2016, researchers from Laurentian University in Ontario, Canada in cooperation with the Ministry of Health in Guatemala and with researchers in Mexico tested and released a design for a low cost trap called an Ovillanta which consists of attractant-laced water in a section of discarded rubber tyre (Gignac, 2016, Roth, 2016). The simple, low-cost trap allows efficient collection and disposal of mosquito eggs and larvae. At regular intervals, the water is run through a filter to remove any deposited eggs and larva. The water, which contains an oviposition pheromone deposited during egg-laying, is reused to attract more mosquitoes. Two studies have shown that this type of trap can attract about seven times as many mosquito eggs as a conventional ovitrap.

Also another process of achieving sustainable larval mosquito control in an eco-friendly manner is the use of trap larvae. This is achieved by providing artificial breeding grounds utilizing common household utensils and destroying larvae by non-hazardous natural means. These include throwing them in dry places or feeding them to larvae eating fishes like *Gambusia affinis*, or suffocating them by spreading a thin plastic sheet over the entire water surface to block atmospheric air (Tietze et al., 1991). Shifting the water with larvae to another vessel and pouring a few drops of kerosene oil or larvicide in it is another option for killing wrigglers.

This method is not preferred due to its environmental impact. Another option is the use of oil drip can or oil drip barrel, which is a common and non-toxic anti-mosquito measure. The thin layer of oil on top of the water prevents mosquito breeding in two ways: mosquito larvae in the water cannot penetrate the oil film with their breathing tube and so are drowned and died; also adult mosquitoes do not lay eggs on the oiled water.

Adult mosquitoes are also targeted using traps. Some newer adult mosquito traps or known mosquito attractants emit a plume of carbon dioxide together with other mosquito attractants such as sugary scents, lactic acid, octenol, warmth, water vapor and sounds (NEA, 2018). By mimicking a mammal's scent and outputs, the trap draws female adult mosquitoes towards it. They are eventually sucked into a net or holder by an electric fan where they are collected. The trap will eventually kill some mosquitoes but their effectiveness in any particular case will depend on a number of factors such as the size and species of the mosquito population, the type and location of the breeding habitat. This method is useful in specimen collection studies to determine the types of mosquitoes prevalent in an area but are far too inefficient to be useful in reducing mosquito populations.

Source Reduction

This method involves reduction or elimination of the breeding habitat of mosquitoes through environmental manipulation. Mosquito breeding grounds can be eliminated at home by removing unused plastic pools, old tyres, or buckets; by clearing clogged gutters and repairing leaks around faucets; by regularly (at most every 4 days) changing water in bird baths; and by filling or draining puddles, swampy areas, and tree stumps. Eliminating such mosquito breeding areas can be an extremely effective and permanent way to reduce mosquito populations without resorting to insecticides. However, this may not be possible in parts of the developing world where water cannot be readily replaced due to irregular water supply.

Another method involves open water marsh management (OWMM), which is the use of shallow ditches to create a network of water flow within marshes and to connect the marsh to a pond or canal. The network of ditches drains the mosquito habitat and lets in fish which will feed on mosquito larvae. This reduces the need for other control methods such as pesticides. Simply giving the predators' access to the mosquito larvae can result in long-term mosquito control. Open-water marsh management is used on both the eastern and western coasts of the United States (Wolfe, 1996).

Also, rotational impoundment management (RIM) is another method that involves the use of large pumps and culverts with gates to control the water level within an impounded marsh. RIM allows mosquito control to occur while still permitting the marsh to function in a state as close to its natural condition as possible (Poulakis et al., 2002). Marshes and swamps especially those with sloping margins allowing fluctuations in the water line, are often very attractive to *Aedes* mosquito which lay their eggs on damp mud and leaf litter (Youdeowei and Service, 1995).

Water is pumped into the marsh in the late spring and summer to prevent the female mosquito from laying her eggs on the soil. The marsh is allowed to drain in the fall, winter, and early spring. Gates in the culverts are used to permit fish, crustaceans, and other marsh organisms to enter and exit the marsh. RIM allows the mosquito-control goals to be achieved while at the same time reducing the need for pesticide use within the marsh.

Integrated Vector Management

Integrated vector management (IVM) is the use of the most environmentally appropriate method or combination of methods to control vector populations. Typical mosquito-control programs using IVM first conduct surveys in order to determine the species composition, relative abundance and seasonal distribution of adult and larval mosquitoes, and only then is a control strategy defined (WHO, 2016). During the past two decades, integrated vector management (IVM) for control of arboviruses has gradually replaced the use of individual vector control strategies and promotes the development of control strategies based on vector biology. IVM also focuses on rigorous monitoring of outcomes, identification and correction of problems encountered in addition to transient suppression of vector populations through interventions (WHO, 2004). Maintaining the positive outcomes of vector control during inter-epidemic periods is a critical factor in integrated vector control (Achee et al., 2015).

Another positive aspect of evidence-based vector control programs is the termination of existing strategies that are found to be ineffective and the development of new methods for vector control based on a better understanding of mosquito biology. The development of new techniques often leads to more efficient tools for the control of mosquito vectors for arboviruses.

In Australia, routine control of *Ae. aegypti* includes the use of lethal ovitraps in addition to larval control, indoor residual spraying (IRS) and use of synthetic pyrethroids such as deltamethirn and lambda-cyhalothrin. These combined methods have successfully controlled *Ae. aegypti* and generally dengue in Australia since 2004 (Rapley et al., 2009).

Conclusion

Aedes aegypti mosquito is a species of mosquito in the gnus *Aedes* originally seen in tropical and subtropical regions, but now found on all continents except Antarctica. It is an aggressive daytime bitter that bite and rest either indoor or outdoor. The adult use a wide range of man-made and natural containers, temporary ground pools, temporary salt pools as well as rock pools for egg laying. The eggs of this species can withstand desiccation, surviving in containers without water for several months, making it a highly resilient and difficult to control mosquito. It prefers mammalian hosts and preferentially feed on humans even in the presence of alternative hosts. The mosquito feed multiple times during one gonotrophic cycle, which has implications for disease transmission. *Ae. aegypti* mosquito is known to transmit a number arboviruses (yellow fever, dengu fever, zika virus and chikunguya virus) that have put billions of people at risk for infection, infect hundreds of millions of humans annually, cause millions of deaths and inestimable morbidity each year. The incidence of these diseases has increased rapidly throughout the tropics, Americas and Asian countries. Hence, the need to address these diseases and the critical needs to augment the armamentarium of tools and approaches to control the vectors that transmit these arboviral diseases.

References

Achee, N. L., Gould, F., Perkins, T. A., Reiner, R. C., Jr., Morrison, A. C., Ritchie, S. A., Gubler, D. J., Teyssou, R. and Scott, T. W. A (2015). Critical assessment of vector control for dengue prevention. *PLoS Neglected Tropical Diseases*, 9:e0003655.

Aik, J., Neo, Z. W., Rajarethinam, J., Chio, K., Lam, W. M., Ng, L. C. (2019). The effectiveness of inspections on reported mosquito larval habitats in households: A case-control study. *PLoS Neglected Tropical Diseases*, 13(6): e0007492.

Akiner, M. M., Demirci, B., Babuadze, G., Robert, V. and Schaffner, F. (2016). Spread of the Invasive Mosquitoes *Aedes aegypti* and *Aedes albopictus* in the Black Sea Region Increases Risk of Chikungunya, Dengue, and Zika Outbreaks in Europe. *PLoS Neglected Tropical Diseases*, 10(4): e0004664.

Barret, A. D. and Higgs, S. (2007). Yellow fever: A disease that has yet to be conquered. *Annual Review of Entomology*, 52: 209 - 229.

Barrett AD, Higgs S (2007). Yellow fever: a disease that has yet to be conquered. *Annual Review of Entomology*, 52: 209–29.

Barrett, D. T. and Higgs, S. (2007). Yellow fever: a disease that has yet to be conquered. *Annual Review of Entomology*, 52: 209–229.

Borgherini G1, Poubeau P, Staikowsky F, Lory M, Le Moullec N, Becquart JP, Wengling C, Michault A, Paganin F. (2007). Outbreak of chikungunya on Reunion Island: early clinical and laboratory features in 157 adult patients. *Clin Infect Dis.*, 44(11):1401-7.

Brady, O. J., Gething, P. W., Bhatt, S., Messina, J. P., Brownstein, J. S., Hoen, A. G., Moyes, C. L., Farlow, A. W., Scott, T. W. and Hay, S. I. (2012). Refining the global spatial limits of dengue virus transmission by evidence-based consensus. *PLOS Neglected Tropical Diseases*, 6: e1760.

Brady, O. J., Golding, N., Pigott, D. M., Kraemer, M. U. G., Messina, J. P., Reiner, R. C. Jr, et al., (2012). Global temperature constraints on *Aedes aegypti* and *Ae. albopictus* persistence and competence for dengue virus transmission. *Parasites and Vectors*, 7:338.

Bravo, J. R., Guzman, M. G. and Kouri, G. P. (1987). Why dengue haemorrhagic fever in Cuba? Individual risk factors for dengue haemorrhagic fever/dengue shock syndrome (DHF/DSS). *Trans R Soc Trop Med Hyg.*, 81(5):816-20.

Brown, J. E., Scholte, E., Dik, M., Hartog, W. D., Beeuwkes, J. and Powell, J. R. (2011). *Aedes aegypti* Mosquitoes imported into the Netherlands. *Emerging Infectious Diseases*, 17(12): 2335–2337.

Brown, J. E., Evans, B. R., Zheng, W., Obas, V., Barrera-Martinez, L., Egizi, A., Zhao, H., Caccone, A. and Powell, J. R. (2014). Human impacts have shaped historical and recent evolution in *Aedes aegypti*, the dengue and yellow fever mosquito. *Evolution*, 68(2):514-25.

Bustamante, M. D. and Lord, C. C. (2010). Sources of error in the estimation of mosquito infection rates used to assess risk of arbovirus transmission. *American Journal of Tropical Medicine and Hygiene*, 82: 1172 - 1184.

Campos, G. S., Bandeira, A. C. and Sardi, S. I. (2015). Zika outbreak in Bahia, Brazil. *Emerging Infectious Diseases*, 21(10): 1885–1886.

CDC (2012). *Surveillance and Control of Aedes aegypti and Aedes albopictus in the United States* https://www.cdc.gov/chikungunya/pdfs/surveillance-and-control-of-aedes-aegypti-and-aedes-albopictus-us.pdf. (Accessed 26 January, 2020).

CDC (2017). 2017 *Case Counts in the US. Final data reported to ArboNET.* https://www.cdc.gov/zika/reporting/2017-case-counts.html (accessed 23 January, 2020).

Clements, A. N. (1999). *The biology of mosquitoes*, Volume II. Egg laying. Cabi, Wallingford.

Clemons, A., Mori, A., Haugen, M., Severson, D. W. and Duman-Scheel, M. (2018). Culturing and egg collection of *Aedes aegypti*. *Ecological Evolution*, 8(16): 7835–7848.

da Costa, C. F., dos Passos, R. A., Lima, J. B. P., Roque, R. A., Sampaio, V. S., Campolina, T. B., Secundino, N. F. C. and Pimenta, P. F. P. (2017). Transovarial transmission of DENV in *Aedes aegypti* in the Amazon basin: a local model of xenomonitoring. *Parasites and Vectors*, 10: 249.

Das, B. Ghosa, S. and Mohanty, S. (2018). Aedes: What Do We Know about Them and What Can They Transmit? *Vectors and Vector-Borne Zoonotic Diseases.* Available from: https://www.intechopen.com/books/vectors-and-vector-borne-zoonotic-diseases/aedes-what-do-we-know-about-them-and-what-can-they-transmit- (Rerieved 28 January, 2020).

de Andrade, P. P., Aragão, F. J. L., Colli, W., Dellagostin, O. A., Finardi-Filho, F., Hirata, M. H., Lira-Neto, A, C., de Melo, M. A., Nepomuceno, A. L., da Nóbrega, F. G., de Sousa, G. D., Valicente, F.

H. and Zanettini, M. H. B. (2016). Use of transgenic *Aedes aegypti* in Brazil: risk perception and assessment. *Bulletin of the World Health Organization*, 94:766-771.

ECDC (2016). Zika virus disease epidemic: Preparedness planning guide for diseases transmitted by *Aedes aegypti* and *Aedes albopictus*. European Centre for Disease Prevention and Control. *Guidelines for surveillance of invasive mosquitoes in Europe.* Available from: http://ecdc.europa.eu/en/healthtopics/vectors/vectormaps/Pages/VBORNET_maps.aspx (Accessed 24 January, 2020).

ECDC (2018). *Aedes aegypti* - current known distribution. *European Centre for Disease Prevention and Control and European Food Safety Authority*. Available from: https://ecdc.europa.eu/en/disease-vectors/surveillance-and-disease-data/mosquito-maps (accessed 26 January, 2020).

ECDC (2012a). *Guidelines for the surveillance of invasive mosquitoes in Europe. European Centre for Disease Prevention and Control (ECDC),* Stockholm. Available from: https://ecdc.europa.eu/sites/portal/files/media/en/publications/Publications/TER-Mosquito-surveillance-guidelines.pdf (accessed 26 January, 2020).

ECDE (2012b). The climatic suitability for dengue transmission in continental Europe. European Center for Disease Prevention and Control. An Agency of the European Union, *Technical Report*, 26 July, 2012.

Eritja, R., Escosa, R., Lucientes, J., Marques, E., Molina, R., Roiz, D. and Ruiz, S. (2005). Worldwide invasion of vector mosquitoes: Present European distribution and challenges for Spain. *Issues in Bioinvasion Science*, 7 (1): 87-97.

Evans, H. C., Elliot, S. L. and Barreto, R. W. (2018). Entomopathogenic fungi and their potential for the management of *Aedes aegypti* (Diptera: Culicidae) in the Americas. *Mem. Inst. Oswaldo Cruz.*, 113 (3).

Fagbami, A. H. (1979). Zika virus infectionin Nigeria: virology and seroepidemological investigation in Oyo state. *Journal of Tropical Medicine and Hygiene.* 83: 213-219.

Ferede, G., Tiruneh, M., Abate, E., Kassa, W. J., Wondimeneh, Y., Damtie, D. and Tessema, B. (2018). Distribution and larval breeding habitats of *Aedes* mosquito species in residential areas of northwest Ethiopia. *Epidemiology and Health*, 40: e2018015.

Foster, W. A., Walker, E. D. (2002). Mosquitoes (Culicidae). In: Mullen G, Durden, L. (editors). *Medical and Veterinary Entomology*. Academic Press. New York. NY, pp245-249.

Franca, R. F., Neves, M. H. L., Ayres, C. F. J., Melo-Neto, O. P. and Brandão Filho, S. P. (2016). First International Workshop on Zika Virus held by Oswaldo Cruz Foundation FIOCRUZ in Northeast Brazil, March 2016, A Meeting Report. *PLoS Neglected Tropical Diseases*, 10 (6): e0004760.

Gignac, L. (2016). Canadian researcher's mosquito trap offers hope in fight against Zika spread. *The Globe and Mail*, Friday, Apr. 08, 2016.

Giovanni, Benelli, Claire L. Jeffries and Thomas Walker (2016). Biological Control of Mosquito Vectors: Past, Present, and Future. *Insects*, 7(4): 52.

Gubler, D. J. (1997). Dengue and dengue hemorrhagic fever: its history and resurgence as a global public health problem. Gubler D, Kuno G, eds. *Dengue and Dengue Hemorrhagic Fever* London, United Kingdom: CAB International, 1–22.

Gubler, D. J. (1998). Dengue and dengue haemorrhagic fever. *Clinical Microbiology Reviews*, 11 (3): 480 - 496.

Gubler, D. J. (2004). The changing epidemiology of yellow fever and dengue, 1900 to 2003: full circle? *Comp. Immunol Microbiol Infect Dis* 27: 319–330.

Hamon, J. (1973). *The Aedes problem in Africa, with particular reference to West Africa.* Office de la Recherche Scientifique et Technique d'Outre-Mer (ORSTOM) France, 5 pp.

Hayes, E. B. (2009). Zika Virus outside Africa. *Emerging Infectious Diseases*, 15 (9): 1347-1350. http://www.who.int/emergencies/zika-virus/situation-report/6Oct2016.pdf?ua=1 (accessed 27 January, 2020).

Huang, Y. M. (2004). The subgenus *Stegomyia* of *Aedes* in the Afrotropical Region with keys to the species (Diptera: Culicidae). Magnolia Press, Auckland, New Zealand. *Zootaxa*, 700: 1 - 120.

Huang, Y. M., Mathis, W. N. and Wilkerson, R. C. (2010). Coetzeemyia, a new subgenus of *Aedes* and a redescription of the holotype female of *Aedes* (*Coetzeemyia*) fryeri (Theobald) (Diptera: Culicidae). Magnolia Press New Zealand. *Zootaxa*, 2638: 24.

Jansen, C. C. and Beebe, N. W. (2010). The Dengue vector, *Aedes aegypti*: what comes next? *Microbes and Infection*, xx: 1-8.

Kotsakiozi, P., Evans, B. R., Gloria-Soria, A., Kamganga, B., Mayanja, M., Lutwama, J., Le Goff, G. Ayala, D., Paupy, C., Badolo, A., Pinto, J., Sousa, C. A. and Powell, J. R. (2019). Population structure of a vector of human diseases: *Aedes aegypti* in its ancestral range, Africa. *Public Health*, 12 June 2019 | https://doi.org/10.3389/fpubh. 2019.00148. (Retrieved 28 January, 2020).

Kramer, V. L., Garcia, R. and Colwell, A. E. (1988). An evaluation of *Gambusia affinis* and *Bacillus thuringiensis* var. *israelensis* as mosquito control agents in California wild rice fields. *Journal of American Mosquito Control Association*, 4:470–478.

Lippi, C. A., Stewart-Ibarra, A. M., Loor, M. E. F. B., Zambrano, J. E. D., Lopez, N. A. E., Blackburn, J. K., et al., (2019). Geographic shifts in *Aedes aegypti* habitat suitability in Ecuador using larval surveillance data and ecological niche modeling: Implications of climate change for public health vector control. *PLoS Neglected Tropical Diseases*, 13(4): e0007322.

Long, S. A, Susan P. Jacups, S. P. and Ritchie, S. A. (2015). Lethal ovitrap deployment for *Aedes aegypti* control: potential implications for non-target organisms. *Journal of Vector Ecology*, 40(1):139-145.

MacNamara, F. N. (1954). Zika virus: A report on three cases of human infection during an epidemic of jaundice in Nigeria. *Trans. R. Soc. Trop. Med. Hyg.*, 48: 139 - 145.

Malone, R. W., Homan, J., Callahan, M. V., Glasspool-Malone, J., Damodaran, L., Schneider, A. D. B., Zimler, R., Talton, J., Cobb, R. R., Ruzic, I. and Smith-Gagen, J. (2016). Zika virus: Medical countermeasure development challenges. *PLoS Neglected Tropical Diseases*, 10 (3): e0004530.

Mayer, S. V., Tesh, R. B. and Vasilakis, N. (2017). The emergence of arthropod-borne viral diseases: a global perspective on dengue, chikungunya and Zika fevers. *Acta Trop.* 166: 155–163.

Morris, A. C., Eggleston, P. and Crampton, J. M. (1989). Genetic transformation of the mosquito *Aedes aegypti* by micro-injection of DNA. *Med. Vet. Entomol.*, 3:1–7.

Natesan, S. K (2019). *Chikungunya Virus*. Medscape Updated: Aug 01, 2019. https://emedicine.medscape.com/article/2225687-overview (Retrieved 27 January, 2020).

NEA (2018). *Effectiveness of Mosquito Traps*. National Environmental Protection Agency, a Singapore Agency Website. https://www.nea.gov.sg/corporate-functions/resources/research/mosquito-traps/effectiveness-of-mosquito-traps (accessed 2 February, 2020).

Nelson, M. J. (1986). *Aedes aegypti: Biology and Ecology*. Pan American Health Organization, Washington, DC, PNSP/86-63, pp. 50.

Norrby, E. (2007). Yellow fever and Max Theiler: The only Nobel Prize for a virus vaccine. *Journal of Experimental Medicine*, 204(12):2779-84.

Nurtop, E., Moyen, N., Dzia-Lepfoundzou, A., Dimi, Y., Ninove, L., Drexler, J. F., Gallian, P., de Lamballerie, X. and Stéphane Priet, S. (2020). *Vector-Borne and Zoonotic Diseases*, 40-42.

Nwankwo, E. N., Okonkwo, N. J., Ozumba, N. A. and Okafor, E. G. (2011). Comparative study on the larvicidal action of Novaluron (Mosquiron 100EC) and *Moringa oliefera* (LAM.) seed oil against *Aedes aegypti* (Diptera: Culicidae) larvae. *African Research Review,* 5(1): 428-436.

Nwankwo, E. N.; Egwuatu, R. I.; Okonkwo, N. J.; Ozumba, N. A; Ewim, S. C; Okolo, K. K. and Akunne, C. (2012a). Toxicity of Novaluron (Mosquiron® 100EC) compared to *Annona muricata* seed oil extract against the larvae of *Aedes aegypti. Asian Journal of Microbiology Biotechnology and Environmental Science*, 14(4): 481-488.

Nwankwo, E. N., Egwuatu, R. I., Okonkwo, N. J., Ozumba, N. A., Ewuim, S. C., Okolo, K. K. and Akunne, C. E. (2012b). Toxicity of Novaluron (Mosquiron 100 EC) compared to *Annona muricata* seed oil extract against the larvae of *Aedes aegypti. Asian Journal of Microbiology, Biotechnology and Environmental Science, 14(4):*481-488.

Nwankwo, E. N., Okonkwo, N. J., Ogbonna, C. U., Akpom, C. J. O., Egbuche, C. M. and Ukonze, B. C. (2015). *Moringa oliefera* and *Annona muricata* seed oil extracts as biopesticides against the second and fourth larval instar of *Aedes aegypti* L. (Diptera: Culicidae). *Journal of Biopesticides*, 8(1): 56-61.

Nwankwo, E. N., Ogbonna, C. U. Ononye, I. P, Ezihe, E. K., Nwude, C. O. and Nwangwu, U. C (2019). Insecticide susceptibility status of *Aedes aegypti* and *Aedes albopictus* in Awka South, Local Government Area, Anambra State, Nigeria. *Entomology, Ornitology and Herpetology,* 8(1): 1-5

Okonkwo, N. J.;Nwankwo, E. N. and Uko, I., Okafor E. G, Ukonze, B. C., Ogbonna, C. U., Egbuche, C. M., Chikezie, F. M. and Obiechina, I. O. (2014). Evaluation of *Moringa oliefera* (Lam) (Moringaceae) seed oil for larval control of *Aedes aegypti* L. (Diptera: Culicidae). *The International Journal of Science and Technology*, 2(12): 75-81

Otu, A. A, Ubong, A. U., Ita, O. I., Hicks, J, P. Egbe, W. O. and Walley, J. (2019). A cross-sectional survey on the seroprevalence of dengue fever in febrile patients attending health facilities in Cross River State, Nigeria. *PLOS ONE.*

Paty, M. C., Six, C., Charlet, F., Heuze, G. Cochet, A., Wiegandt, A., Chappert, J. L., Dejour-Salamanca, D., Guinard, A., Soler, P., Servas, V., Vivier-Darrigol, M., Ledrans, M., Debruyne, M., Schaal, O., Jeannin, C., Helynck, B., Leparc-Goffart, I. and Coignard, B. (2014). Large number of imported chikungunya cases in mainland France, 2014: a challenge for surveillance and response. *Euro Surveill.*, 19(28):20854.

Petersen, E. 1., Wilson, M. E., Touch, S., McCloskey, B., Mwaba, P., Bates, M., Dar, O., Mattes, F., Kidd, M., Ippolito, G., Azhar, E. I. and Zumla, A. (2016). Rapid Spread of Zika Virus in the Americas--

Implications for Public Health Preparedness for Mass Gatherings at the 2016 Brazil Olympic Games. *WHO 2016 Zika situation report*. 31 March 2016 https://www.who.int/emergencies/zika-virus/situation-report/31-march-2016/en/ (Accessed 27 January, 2020).

Phuc, H. K., Andreasen, M. H., Burton, R. S., Vass, C., Epton, M. J., Pape, G., Fu, G., Condon, K. C., Scaife, S., Donnelly, C. A., et al., (2007). Late-acting dominant lethal genetic systems and mosquito control. *BMC Biol*., 5:11–21.

Poulakis, G. R., Shenker, J. M. and Taylor, D. S. (2002). Habitat use by fishes after tidal reconnection of an impounded estuarine wetland in the Indian River Lagoon, Florida (USA). *Wetlands Ecology and Management*, 10 (1): 51-69.

Powell, J. R. (2018). Mosquito-Borne Human Viral Diseases: Why *Aedes aegypti*? *American Journal of Tropical Medicine and Hygiene*, 98(6): 1563–1565.

Powell, J. R. and Tabachnick, W. J. (2013). History of domestication and spread of *Aedes aegypti* -a review. *Mem Inst Oswaldo Cruz*., 108 (Suppl 1): 11–17.

Provonsha, A. V. (1983). *Biological notes on mosquitoes*. http://www. mosquitoes.org/Life cycle.html. (Accessed 27th January, 2020).

Rapley, L. P., Johnson, P. H., Williams, C. R., Silcock, R. M., Larkman, M., et al., (2009). A lethal ovitrap-based mass trapping scheme for dengue control in Australia: II. Impact on populations of the mosquito *Aedes aegypti*. *Med. Vet. Entomol*., 23: 303–316.

Rezza G1, Nicoletti L, Angelini R, Romi R, Finarelli AC, Panning M, Cordioli P, Fortuna C, Boros S, Magurano F, Silvi G, Angelini P, Dottori M, Ciufolini MG, Majori GC, Cassone A. (2007). Infection with chikungunya virus in Italy: an outbreak in a temperate region. *Lancet*, 370(9602):1840-6.

Rezza, G. and Weaver, S. C. (2019). Chikungunya as a paradigm for emerging viral diseases: Evaluating disease impact and hurdles to vaccine development. *PLoS Neglected Tropical Diseases*, 13(1): e0006919.

Rodriguez-Cavalloa, E., Guarnizo-Mendeza, J., Yepez-Terrilla, A., Cardenas-Riveroab, A., Diaz-Castillob, F. and Mendez-Cuadroc, D. (2019). Protein carbonylation is a mediator in larvicidal mechanisms of *Tabernaemontana cymosa* ethanolic extract. *Journal of Kinf Saud University-Science*, 31(4): 464-471.

Roth, A. (2016). How Canadian Scientists Plan to Fight Zika With Old Tires and Milk." *Motherboard*, April 7, 2016.

Rwang, P. G., Effoim, O. E., Mercy, K. P. and Etokakpan, A. M. (2016). Effects of *Psidium guajava* (Guava) Extracts on Immature Stage of Mosquito. *International Journal of Complementary and Alternative Medicine*, 4(5): 00132.

Savage, H. M., Ezike, V. I., Nwankwo, C. A. N., Spiegel, R., Miller, B. R. (1992). First record of breeding populations of *Aedes albopictus* in continental Africa: implications for arboviral transmission. *Journal of American Mosquito Control Association*, 8:101–102.

Schaffner, F. and Mathis, A. (2014). Dengue and dengue vectors in the WHO European region: Past, present, and scenarios for the future. *The Lancet Infectious Diseases*, 14:1271-1280.

Scolari, F., Casiraghi, M. and Bonizzoni, M. (2019). *Aedes* spp. and Their Microbiota: A Review. *Frontiers in Microbiology*, 10: 2036.

Scott, T. W. and Takken, W. (2012). Feeding strategies of anthropophilic mosquitoes result in increased risk of pathogen transmission. *Trends in Parasitology*, 28 (3): 114-121.

Service, M. W. (1996). *Medical Entomology for students*. Chapman and Hall, London, UK. pp 1-80.

Sikka, V., Chattu, V. K., Popli, R. K., Galwankar, S. C., Kelkar, D., Sawicki, S. G., Stawicki, S. P. and Papadimos, T. J. (2016). The emergence of Zika virus as a global health security threat: A review and a consensus statement of the INDUSEM Joint working Group (JWG). *Journal of Global Infectious Diseases*, 8 (1): 3-15.

Smith, L. (2016). *Zika virus: Symptoms, Facts and Diagnosis.* http://www.medicalnewstoday.com/articles/305163.php. (Accessed January 26, 2020).

Soumahoro, M., Fontenille, D., Turbelin, C., Pelat, C., Boyd, A., Flahault, A. and Hanslik, T. (2010). Imported Chikungunya Virus Infection. *Emerging Infectious Diseases*, 16(1): 162–163.

Souza-Neto, J. A. and Powell, J. R. (2019). *Aedes aegypti* vector competence studies: A review. Infection, *Genetics and Evolution*, 67: 191-209.

Steinwascher, K. (2018). Competition among *Aedes aegypti* larvae. *PLoS ONE*, 13(11): e0202455.

Thomas, M. B. (2017). Biological control of human disease vectors: a perspective on challenges and opportunities. *Biocontrol*, 63(1): 61–69.

Tietze, N. S., Hester, P. G., Hallmon, C. F., Olson, M. A. and Shaffer, K. R. (1991). Acute toxicity of mosquitocidal compounds to young mosquitofish, *Gambusia affinis*. *Journal of American Mosquito Control Association*, 7:290–293.

Tomori, O. (2004). Yellow fever: the recurring plague. *Crit Rev Clin Lab Sci.*, 41 (4): 391–427.

Trpis M (1972) Breeding of *Aedes aegypti* and *Ae. simpsoni* under the escarpment of the Tanzanian plateau. *Bulletin of World Health Organanisation*, 47: 77–82.

Vainio, J. and Cutts, F. (1998). *Yellow fever*. World Health Organisation. Geneva Switzerland. WHO/EPI/GEN/98, 11: 1 - 87.

Vidyadhara, R. P and Naveen K. S. (2018). Evaluation of thrombocytopenia in dengue infection along with seasonal variation. *IAIM*, 5(2): 57-63.

Vorou, R. (2016). Zika virus, vectors, reservoirs, amplifying hosts, and their potential to spread worldwide: what we know and what we should investigate urgently. *International Journal of Infectious Diseases*, 48:85-90.

WASH (2016). *Aedes aegypti* vector control and prevention measures in the context of Zika, Yellow Fever, Dengue or Chikungunya. Technical Guidance Version. *WCA Regional Group* https://www. humanitarian response.info/en/operations/west-and-central-africa/watersanitation-hygiene (accessed 25 January, 2020).

Weaver, S. C. and Lecuit, M. (2015). Chikungunya virus and the global spread of a mosquito-borne disease. *N. Engl. J. Med.*, 372: 1231–1239.

Weaver, S. C. and Reisen, W. K. (2010). Present and future arboviral threats. *Antiviral Research*, 85 (2): 328 - 345.

Weetman, D., Kamgang, B., Badolo, A., Moyes, C. L., Shearer, F. M., Coulibaly, M. et al., (2018). *Aedes* mosquitoes and aedes-borne arboviruses in Africa: Current and future threats. *International Journal of Environmental Research and Public Health*, 15:220.

WHO (1995). *Report of the consultation on key issues in dengue vector control towards the operationalization of a global strategy.* World Health Organisation CTD/FIL(DEN)/IC/96.1: Geneva, Switzerland, pp 48.

WHO (2000). *Report of the informal consultation on: Strengthening implementation of the global strategy for Dengue fever/Dengue Haemorrhagic fever prevention and control.* World Health Organisation. WHO/CDS/ (DEN)/IC/2000.1: Geneva, Switzerland, pp 20.

WHO (2004). *Global Strategic Framework for Integrated Vector Management.* World Health Organisation; Geneva, Switzerland.

WHO (2005). *The yellow fever situation in Africa and South America in 2004.* World Health Organisation Weekly Epidemiological Record, 80: 250-256.

WHO (2006). *Pesticides and Their Application.* World Health Organisation; Geneva, Switzerland.

WHO (2009). *Dengue: Guidelines for Diagnosis, Treatment, Prevention and Control.* World Health Organisation. WHO/CDS/(DEN)/ IC/2000.1: Geneva, Switzerland, pp 20.

WHO (2012a). *Global Strategy for Dengue Prevention and Control 2012 – 2020.* World Health Organisation Geneva, Switzerland. P 35.

WHO (2012b). *Yellow fever in Ghana.* Report from World Health Organisation.

WHO (2014). *Yellow fever Fact sheet N°100."* World Health Organization. Archived from the original on 19 February 2014. Retrieved 27 January 2020.

WHO (2015). *Core vector control methods.* World Health Organization http://www.who.int/malaria/areas/vector_control/core_methods/en/ (Retrieved January, 2020).

WHO (2016a). *Countries and territories reporting mosquito-borne Zika virus transmission.* World Health Organisation, Geneva.

WHO (2016b). *Integrated Vector Management.* World Health Organisation http://www.who.int/neglected_diseases/vector_ecology/ivm_concept/en/ (Retrieved January, 2020).

WHO (2017a). *Global vector control response 2017-2030.* Geneva: WHO and UNICEF.

WHO (2017b). Yellow fever. W*orld Health Organization Media Centre Fact sheets*.

WHO (2019). *Dengue and severe dengue.* https://www.who.int/news-room/fact-sheets/detail/dengue-and-severe-dengue. (Retrieved 28 January, 2020).

WHOPES (2011). *Global insecticide use for vector-borne disease control: a 10-year assessment* (2000–2009), 5th ed. Geneva, World Health Organisation.

Wikan, N. and Smith, D. R. (2017). First published report of Zika virus infection in people: Simpson, not MacNamara. *The Lancet Infectious Diseases*, 17(1): 15-17.

Wilson AL, Courtenay O, Kelly-Hope LA, Scott TW, Takken W, Torr SJ, et al., (2020) The importance of vector control for the control and elimination of vector-borne diseases. *PLoS Neglected Tropical Diseases*, 14(1): e0007831.

Wolfe, R. J. (1996). Effects of open marsh water management on selected tidal marsh resources: a review. *Journal of the American Mosquito Control Association*, 12 (4):701-712.

Yan-Jang S. Huang, Stephen Higgs and Dana L. Vanlandingham Insects (2017). *Biological Control Strategies for Mosquito Vectors of Arboviruses*. Mar., 8(1): 21.

Yasuno, M. and Tonn, R. J. (1970). Bionomics of *Toxorhynchites splendens* in the larval habitats of *Aedes aegypti* in Bangkok, Thailand. *Bulletin of World Health Organisation*, 43:762.

Youdeowei, A. and Service, M. W. (1995). *Pest and vector management in the tropics*. Longman Nigeria Plc. 399pp.

In: *Aedes aegypti*
Editor: Valèrie Megens

ISBN: 978-1-53618-197-5

Chapter 3

ENTOMOPATHOGENIC FUNGI: APPLICATIONS TO THE BIOLOGICAL CONTROL OF CULICIDAE VECTORS OF DISEASE

***Márcia Jordana F. Macêdo*[1], *Nara Juliana S. Araújo*[1], *Luis Pereira-de-Morais*[1], *Yedda Maria Lobo S. Matos*[2], *Henrique Douglas M. Coutinho*[2,*], *Francisco Assis B. Cunha*[3] *and Maria Flaviana B. Morais-Braga*[4]**

[1]Program of Residency in Collective Health
[2]Laboratory of Microbiology and Molecular Biology
[3]Laboratory of Bioprospection of the Semiarid
[4]Laboratory of Applied Mycology of Cariri
Regional University of Cariri, Crato, Ceará, Brazil

[*] Corresponding Author's E-mail: hdmcoutinho@gmail.com.

ABSTRACT

In recent years, the use of entomopathogenic fungi has been a promising tool for the biocontrol of Culicides that threaten public health. We systematically reviewed (2007-2018), studies that evaluated the main entomopathogenic genera used in the biological control of these vectors. Among the genera reported in this review, *Metarhizium* and *Beauveria* stand out as the best investigated over the years, being considered highly virulent, mainly, against species of the genera *Aedes* and *Anopheles*, in their larval and adult stages. While *Culex* spp. remains little researched. From the remaining entomopathogens, the oomycete *Leptolegnia chapmanii*, appears as a promising candidate in the infectivity and death of *Ae. aegypti*.

Keywords: entomopathogenic fungi, vectors, *Aedes aegypti*

INTRODUCTION

Resistance to chemical insecticides is increasingly affecting public health and constitutes a major threat to the control and prevention of diseases such as dengue, zika, chikungunya, yellow fever, malaria, filariasis, among others [1, 2]. The Culicidae family stands out among the mosquito vectors. This family, in turn, has three Subfamilies: Toxorhynchitinae, which has no importance in pathogen transmission; Anophelinae and Culicinae, which continue to have a significant impact on human health, since they contain a large number of medically relevant insects, such as species from *Aedes*, *Culex* and *Anopheles* genera [3].

The emerging resistance to chemical insecticides in these Culicides points to an urgent need for alternative tools which can be used to control these vectors [4]. One method that is currently being explored and which has gained interest is the use of bio-insecticides, such as entomopathogenic fungi [5].

The advantages of using bio-insecticides include their biodegradability and target specificity [6], in addition to being ubiquitous representatives and relatively cheap for mass reproduction [7, 8]. These can cause substantial mortality in Culicidae larvae with less than 24 hours of vector exposure to these pathogens, depending on the lineage, formulation and concentration used, among other aspects [9, 10]. In addition to their infection capacity, the use of these fungi has also been aimed at determining their safety for public health [11, 12].

Entomopathogenic fungi are grouped mainly within Ascomycota, Basidiomycota and Zygomycota phyla, although some species also belong to Chytridiomycota [13]. Within these, Ascomycota houses the most important genera, such as *Beauveria* and *Metarhizium*, which are considered extremely effective against disease-bearing Culicidae [14-16]. Within Zygomycota phylum, the most entomopathogenic species are in the Entomophthorales order [17], with only a small number of representatives from Basidiomycota being relevant for biological control [13].

Beauveria is one of the entomogenic genera with worldwide distribution, widely investigated as a biological tool against disease vectors [18-20]. *B. bassiana* stands out among the species that make up this genus, being highly effective in *Ae. aegypti*, *Cx. quinquefasciatus* and *Anopheles* spp., larval stage control [5, 21, 22], whose species are susceptible to infections by this pathogen. Studies have also shown their potential in the adult vector control of these species, such as reduced longevity [23, 24], blood supply and fecundity [25, 26], as well as reduced insecticide resistance expression [27-29].

Metarhizium, is an entomogenous fungi like *Beauveria*, whose insect pathogenesis has been extensively documented [30]. *Metarhizium* has a cosmopolitan distribution [31], with the genus gathering several species, with *M. anisopliae* being the research focus in mosquito control [32-34]. This species is hemibiotrophic [35], presenting a range of hosts and encompassing five insect orders [31], within which the fungus stands out given its potential against arbovirus vectors and other diseases present in the Diptera order [36-38].

The aforementioned entomopathogens exhibit several stages in the development of fungal infections in adult insects and present a similar process to other entomopathogenic fungi [13], ranging from spore fixation to mosquito cuticle, followed by germination, cuticle penetration and internal dissemination [13]. These end up invading the haemocle of the entire host, exploiting nutritional resources, killing the hosts by starvation, dehydration and toxemia [31]. On the other hand, their insect mechanism of infection during the larval life stage can begin when the fungus contacts, germinates and penetrates the respiratory tract, blocking the respiratory mechanism [39]. Larval killing effectiveness is strongly influenced by temperature, a factor which modulates larval mortality rate, being indispensable for germination and fungal sporulation [40].

In view of the above, given the importance of entomopathogenic fungi as a biological tool for the control of mosquitoes affecting public health, the objective of this review was to collect and update information surrounding the main genera used in Culicidae disease vector biological control. The results will be presented in subsections, with each subsection gathering information on the most commonly used entomopathogens in Culicidae control. The main methods employed, mechanisms of action, their effectiveness as bio-insecticidal agents, as well as challenges which still exist in the use of these fungi and future perspectives, will be highlighted.

METHODOLOGY

This research consists of a systematic literature review on the main entomopathogenic fungi genera, demonstrating their importance in the biological control of Culicidae vectors of diseases.

The bibliography used and cited was carefully researched and compiled, in important databases, namely Scopus, PubMed, BioMed, Bioline, Doaj and Scielo. The papers cited throughout the article which are not referring to results covered a period between 1998 to 2018. The search for scientific information on the subject of this review comprised a period between 2007-2018. Word combinations such as "entomopathogenic fungi + *Aedes*," "entomopathogenic fungi + *Anopheles*" and "entomopathogenic fungi + *Culex*," "entomopathogenic fungi + Culicidae," were used as search terms.

Entomopathogenic Fungi in the Biocontrol of Culicidae

A total of 204 articles which met the inclusion criteria and were within the limits of the time period were compiled. Of the 204 articles analyzed, 140 addressed the main entomopathogenic genera tested in Culicidae control, with applications in the laboratory and in environmental conditions. Other articles presented general information regarding their mechanism of action, general characteristics and evolution perspectives on the entomopathogenicity of these fungi as bioinseticidal agents.

In this review, 10 genera and 27 entomopathogenic fungal species that have already been tested and show effectiveness against vectors of *Aedes*, *Anopheles* and *Culex* genera were identified. Members of Ascomycota phylum, such as *Metarhizium*, appeared in 41.4% of publications, followed by *Beauveria* with 35%, which confirms these genera are the most investigated in the biocontrol of the aforementioned target vectors over the years (2007-2018). Other fungi genera, such as *Culicinomyces*, *Isaria*, *Lecanicillium*, *Verticilium*, *Aspergillus*, *Fusarium* and oomycetes such as *Lagenidium* and *Leptolegnia* collectively account for 23.6% of publications.

Metarhizium Genus

Metarhizium genus is among the most important entomopathogenic fungi in biological control. Among the species described, are *M. anisopliae, M. brunneum* and *M. robertsii* [34]. These can be highly virulent for a wide range of insects that threaten public health, including *Aedes*, *Culex* and *Anopheles* genera [41]. On the other hand, other species such as *M. acridum* and *M. album* are more specific for certain insects, such as orthopterans or hemiptera [34]. Similarly, *M. majus* is more closely associated with Coleoptera, in agricultural pest control [42]. However, these species have also been tested against mosquito disease vectors [43].

Of the 10 species identified in this review, *M. anisopliae* is well documented with potential applicability, especially against Culicidae from the *Aedes* and *Anopheles* genera in their larval and adult stages, with *Ae. aegypti*, *An. stephensi* and *An. gambiae* being the most studied target species (Table 1).

Species from *Metarhizium* genus are usually found in the soil with their pathogenic mechanisms of action occurring through their conidia or blastospores, the latter being considered more virulent [41] since they have a thinner cell wall, which may facilitate and accelerate the infectious process on the insect. *M. anisopliae*, presents a great potential for field application, as demonstrated by recent studies [44-48].

Metarhizium action has been studied in many insect species. Currently, it has been highlighted as a biological control strategy for mosquito populations with sanitary importance in several countries, as shown by studies conducted in Brazil [37, 47, 49-50], USA [51, 52], China [53, 54], Australia [26], Mexico [33, 42, 55-56], Africa [44, 57, 58], India [59], among others. These studies have tested *Metarhizium* applicability in both laboratory and environmental conditions. Results on the use of these fungi as biological tools against Culicidae disease vectors have been promising, when compared to the excessive use of synthetic insecticides. Currently, the constant use of a limited number of chemical pesticides has caused a major problem in the integrated management of mosquito vectors, given the emergence of resistant forms. In addition, it has also affected human health and the environment, causing environmental pollution and affecting non-target organisms [5].

Diseases transmitted by vector mosquitoes have been increasing significantly each year, including arboviruses such as dengue, Zika, Chikungunya and urban yellow fever, which are transmitted by *Ae. aegypti* vector [20]. Another Culicide capable of transmitting some of these viral diseases is *Ae. albopictus*, whose larvae have also been explored to evaluate *Metarhizium* larval activity [32, 60], which despite displaying potential, the authors state the fungus takes a long time to kill 50% of the population, when compared to synthetic insecticides [32]. Moreover, further studies need to be extended to field conditions.

Table 1. Species of the genus *Metarhizium* used in the biological control of Culicidae

Species	**Target Culicide**	**Target stage of life**	**Application methods**	**Concentrations of conidia and/or blastospores**	**References**
M. anisopliae	*An. stephensi*	Larvae, Pupae and Adults	Laboratory	1 x 10^2 - 5 x 10^{10} /mL	[36], [52] and [85]
M. anisopliae	*An. stephensi* and *An. gambiae*	Larvae	Laboratory and Field	4.7 x 10^8 /cm^2	[45] and [75]
M. anisopliae	*An. arabiensis*	Adults	Field	3.9 x 10^{10} /m^2	[44]
M. anisopliae	*Ae. aegypti*	Eggs	Laboratory	5 x 10^6 /cm^2	[49], [67] and [68]
M. anisopliae	*Ae. aegypti*	Adults	Laboratory and Field	5 x 10^6 - 5.96 x 10^7 /cm^2	[14] and [56]
M. anisopliae	*Ae. aegypti*	Larvae	Laboratory	1 x 10^5 - 10^9 /mL^{-1}	[61], [62] and [65]
M. anisopliae	*Ae. aegypti*, *An. stephensi* and *Cx. quinquefasciatus*	Larvae	Laboratory	1 x 10^5 - 10^8 /mL^{-1}	[9]
M. anisopliae	*Ae. aegypti*	Adults	Laboratory	1 x 10^6 - 10^9 /mL^{-1}	[37], [66] and [130]
M. anisopliae	*Ae. aegypti*	Adults	Laboratory	1.6 x 10^8 - 6 x 10^8/mL^{-1}	[33] and [55]
M. anisopliae	*An. funestus*	Eggs, larvae and Adults	Laboratory	--	[73]
M. anisopliae	*An. funestus* and *An. gambiae*	Adults	Field	1 x 10^{10} - 4 x 10^{10} /m^2	[77]
M. anisopliae	*An. gambiae*	Adults	Laboratory and Field	2 x 10^{10} /m^{-2}	[57] and [72]
M. anisopliae	*Ae. aegypti*	Adults	Field	1 x 10^8 - 10^9 /mL^{-1}	[46], [47] and [48]
M. anisopliae	*Ae. aegypti*	Larvae and Adults	Laboratory	8.93 x 10^5 /mL	[131]
M. anisopliae	*Ae. aegypti*	Larvae	Laboratory and Field	1 x 10^8 - 5 x 10^6 /mL^{-1}	[59] and [101]
M. anisopliae	*Ae. albopictus*	Larvae	Laboratory	1 x 10^4 - 10^8 /mL	[32] and [60]
M. anisopliae	*Ae. aegypti* and *Ae. albopictus*	Adults	Laboratory	1.6 x 10^{10} /m^2	[132]

Table 1. (Continued)

Species	Target Culicide	Target stage of life	Application methods	Concentrations of conidia and/or blastospores	References
M. anisopliae	*An. gambiae* and *An. arabiensis*	Eggs	Laboratory	5 x 10^6 - 10^7/cm^2	[84]
M. anisopliae	*Ae. aegypti*	Adults	Field	1 x 10^8/cm^2	[50]
M. anisopliae	*Cx. pipiens*	Larvae	Laboratory and Field	1 x 10^8 - 4 x 10^8 /mL	[87] and [88]
M. anisopliae	*An. albimanus*	Larvae and Adults	Laboratory	2.6 x 10^7 - 2.7 x 10^8/mL	[15]
M. anisopliae	*Ae. aegypti*	Adults	Laboratory	3.3 x 10^5 /cm^{-2}	[38]
M. anisopliae	*Ae. aegypti*	Eggs	Laboratory	3.3 x 10^3 - 10^5 /g	[70]
M. anisopliae	*An. arabiensis*	Adults	Field	8 x 10^{10}/m^2	[79]
M. anisopliae	*Ae. aegypti*, *An. stephensi* and *Cx. quinquefasiatus*	Larvae	Laboratory	3.48 x 10^3/mL	[86]
M. anisopliae	*Ae. aegypti*	Eggs	Laboratory	2.8 x 10^2 - 3 x 10^5 /cm^{-2}	[69] and [71]
M. anisopliae	*An. stephensi*	Adults	Laboratory	1 x 10^6 - 10^7 and 2 x 10^6/cm^2	[82]
M. anisopliae	*An. gambiae*	Adults	Laboratory and Field	1 x 10^{11} - 3 x 10^{11} /m^2	[97] and [110]
M. anisopliae	*An. gambiae* and *An. arabiensis*	Adults	Field	2 x 10^{10} - 8 x 10^{10} /m^2	[78]
M. anisopliae	*An. gambiae*, *arabiensis* and *An. funestus*	Adults	Laboratory	--	[27]
M. acridum, *M. album*, *M. majus*, *M. lepidiotae*, *M. frigidum* and *M. flavoviride*	*Ae. aegypti*	Adults	Laboratory	--	[43]
M. brunneum	*Cx. quinquefasciatus* and *Ae. aegypti*	Larvae	Laboratory	1 x 10^7 /mL^{-1}	[34] and [64]
M. brunneum	*Ae. aegypti*, *An. stephensi* and *Cx. quinquefasciatus*	Larvae	Laboratory	1 x 10^6 - 10^8 /mL ^{1}	[10] and [133]
M. brunneum and *M. robertsii*	*An. messeae* and *Cx. pipiens*	Larvae	Laboratory	1 x 10^6 - 2 x 10^6 /cm^2	[134]
M. pingshaense	*An. coluzzii* and *An. gambiae*	Adults	Laboratory	1 x 10^6 - 8.0 x 10^7 /mL	[58] and [83]
M. robertsii	*Ae. flavescens*	Larvae	Laboratory	3 x 10^4 - 10^6 /mL	[135]

Several studies on *Ae. aegypti* culicide report the use of *M. anisopliae* entomopathogen as a bio-larvicidal agent for this vector, such as in the study by Butt et al. [61], which showed the fungus was highly effective at killing this vector's larvae, with mortality rates of 60-90% observed at 72-96 hours post inoculation. However, larval death was not the result of a pathogenesis process by traditional cuticle adhesion and penetration means, which further corroborates with a later study carried out by Greenfield et al. [62], who reported *M. anisopliae* infectious pathway is similar to *Culicinomyces clavisporus* aquatic fungal pathogen, which is ingested instead of penetrating the cuticle [63]. Butt et al. [61] presented evidence that larval mortality appears to be linked to autolysis through *Hsp70* gene activity, with this response being modified by protease inhibitors. Results similar to these were also observed by Greenfield et al. [9].

Besides *M. anisopliae*, other species such as *M. brunneum* are also described in experiments by Alkhaibari et al. [64] and Alkhaibari et al. [10], which tested the virulence of *M. brunneum* conidia and blastospores on *Ae. aegypti*. According to the authors, the fact that blastospores have multiple entry routes (cuticle and intestine) may explain why this form of the inoculum kills *Ae. aegypti* in a relatively short time (12-24 hours), which is significantly faster than when the larvae are exposed to conidia. In addition, hydrophilic blastospores produce a copious mucilage, which probably facilitates adhesion to the host [64].

In the literature, synergistic combinations of plant bioactive metabolites and fungi have also been documented. The use of these formulations may increase *M. anisopliae* efficiency when used against *Ae. aegypti* in their larval stage [65]. Gomes et al. [65], used two fungal concentrations (1x 10^7-10^8 conidia/mL^{-1}) in their study suspended in Neem oil (*Azadirachta indica*) at 0.001%, where the survival rate of the larvae exposed to this combination was found to be 36% and 12%, respectively. The authors showed that increasing fungal concentration slightly decreases the survival rate of larvae exposed to fungi and Neem.

The synergism of the combined action of this fungus with other products, in addition to essential oils, such as the insecticide imidacloprid, which has been the subject of previous studies, have also been investigated [37] in order to improve *M. anisopliae* efficiency against *Ae. aegypti*, the effect of which can be increased with ultra low concentrations of the insecticide resulting in higher mortality after relatively short exposure times. It is already known the joint action of these two agents may also decrease vector capacity of this species [37].

For the adulticidal action, several methodologies exist to evaluate the efficiency of this fungus against this vector, among which is the use of sprays [66], conidia impregnation in cloth surfaces [46-48,50], filter paper [37, 55-56] or by immersion [14], in addition to the exposure of mosquitoes to culture media [43] where the fungus grows. All the aforementioned methodologies have been proven to be effective.

With respect to the effects of this fungus against eggs of this vector, the study by Luz et al. [67] was the first to demonstrate *M. anisopliae* efficacy and those of other entomopathogenic fungi against *Ae. aegypti* under laboratory conditions. Similar studies by Luz et al. [49], Albernaz et al. [68] and Santos et al. [69] were also later observed. These showed that conidia formulated in water or oil, when applied to eggs in filter paper and incubated in humidity close to saturation, can inhibit larvae hatching, depending on the incubation time period. Recent discoveries have also shown that *Ae. aegypti* can be infected by *M. anisopliae* when exposed to soil, mixed with conidia and water [70]. If a minimum amount of water is not present at hatching, larvae from this Culinidae cannot survive, regardless of whether or not eggs are infected with fungi [71]. The fact that this fungus is a potential controlling agent for this vector in reproduction places with high humidity is important, since the incidence of arbovirose cases increases concomitantly with the rainy period. The possible utility of the fungus as a mycoinsecticide with applicability in the vector's breeding sites is shown to be promising.

Among other Culicidae, that have also been tested the exposure by *Metarhizium*, are some species from the Anophelinae Subfamily such as *An. stephensi*, *An. gambiae*, *An. arabiensis*, *An. coluzzii*, *An. albimanus*, *An. messeae* and *An. funestus* [9, 15, 36, 44, 45, 51, 52, 57, 58, 72, 73], which are malaria vectors, responsible for one of the greatest health problems in the world [74]. The challenges in malaria control, due to the continuous dependence on spraying with synthetic products, intensified this growing interest in the search for alternative tools, such as the use of these entomopathogenic fungi in the biocontrol of these mosquitoes [74].

Experiments carried out over the years highlight some of the main methods used in laboratory tests and in field conditions during research. Efforts have been made to validate the efficacy of pathogenic fungi, such as those from *Metarhizium* genus on malaria vectors. Bukhari et al. [45] developed formulations to improve *M. anisopliae* spore efficacy and dissemination for the control of anopheline larvae in western Kenya, under field conditions. Adverse effects were observed when the fungus (4.7×10^8 spores/cm^2) was formulated with synthetic oil (ShellSol T), showing uniform spreading and being indicated as a promising vehicle for fungal spores to attack *An. gambiae* and *An. stephensi* larvae. Certain factors that affect the larval mortality of these species, such as larval characteristics (species type, stage and larval density), fungus (species and concentration) and environmental effects (duration of exposure and food availability) have also been studied [75]. Fungi can also be formulated and applied as chemical insecticides [76] and can be delivered through new strategies such as resting targets [57, 77], eaves curtains [78], landing gear [79] or odor bait traps [44].

The method proposed by Mmbando et al. [79], using a solar-powered trap, was tested in Africa for the control of *An. arabiensis* adult population. Mosquitoes were attracted, contaminated and killed when they sought out outdoor hosts, even in the presence of human hosts. The experiment was developed in a semi-field system, where landing boxes (containing natural human odors and carbon dioxide) were sprayed with *M. anisopliae* fungus (8×10^{10} conidia/m^2 formulated with mineral oil). The study also demonstrated the traps can be used in combination with larvicides such as pyriproxyfen, which can sterilize adult mosquitoes [80, 81], and similarly to fungi, also have the ability to promote self-dissemination [79], transporting to aquatic habitats, in order to control larvae populations. In other studies, entomopathogenic fungi have also been reported for their ability to reduce the daily survival of adult anopheline mosquitoes

[36, 73, 82, 83], since fungal conidia are easily disseminated within the mosquitoes' body through the cuticle [44], suppress their cellular defence system, as well as altering the vector's ability to fly, given the pathogen also grows in the legs and wings [36], which allows visual confirmation of fungal growth in mosquito cadavers 8-9 days after exposure [72].

Simple strategies to attract and kill have also been used and have been shown to be promising to infect mosquitoes that come into contact with conidial host-seeking or oviposition behavior [57, 78]. In their studies, Mnyone et al. [57] reported the efficacy of *M. anisopliae* against *An. gambiae* adults, by applying conidia to clay panels (which simulated traditional houses in Tanzania). Similar studies have been developed by Mnyone et al. [78], where the use of eyelash curtains treated with the fungus were used in experimental huts. In previous studies, clay pot impregnation, containing spores, is also reported as satisfactory for fungal application (4 x 10^{10} conidia/m^2) and highly attractive for *Anopheles* mosquitoes [77].

When comparing the ovicidal and pupicidal effects of the available studies with other reported stages (larval and adult) against malarial vectors, these remain scarcely explored [84, 85]. Female *Anopheles* are considered oviposition generalists, however unlike *Ae. aegypti*, their eggs can only withstand short periods outside of the water [84]. The egg stage, which is difficult to eliminate with synthetic insecticides, may provide a potential target for transient habitats with entomopathogenic fungi and fungi-based biological control, since they are susceptible to fungal infections [84].

The applicability of *Metarhizium* against *Culex* genus still remains scarcely investigated, especially in field conditions. *Cx. quinquefasciatus* (Subfamily: Culicinae), the main *Wuchereria bancrofti* vector species in Brazil is the etiological agent of filariasis in humans. The vector has already been tested under laboratory conditions in its larval life stage [9, 34, 64, 86]. However, despite showing significantly reduced survival rates during *M. brunneum* and *M. anisopliae* fungal exposure, the number of studies addressing the potential of the aforementioned species during other life cycle stages still need to increase, especially those addressing its application in the field.

For *Cx. pipiens* species, an important arbovirus vector, including the West Nile virus, biochemical studies indicate *M. anisopliae* application as a larvicidal agent against this vector causes a significant decrease in larval proteins [87], suggesting the toxins secreted by this pathogen damages these proteins, which are essential for the synthesis of chitin, a primordial macromolecule for the life of insects [87]. In the study recently conducted by Shoukat et al. [88], the combined effect of this fungus with synthetic insecticides was investigated under laboratory and environmental conditions, where their mixture proved to be an effective tool for *Cx. pipiens* integrated management.

Although many successful laboratory and environmental evaluations addressing the efficacy of this entomopathogenic fungus have already been carried out, through various formulations, isolate selection, techniques and delivery formats, including production optimization, which is also reported in the literature [89], additional information is still needed to validate its introduction and application to current control programs.

Beauveria Genus

Beauveria genus has been attracting attention as a biocontrol agent. It is a cosmopolitan soil fungus [90] and is generally found in infected insects, both in temperate and tropical areas around the world and Turkey, Cote d'Ivoire, Equatorial West Africa, Central Africa, South Africa, Bahamas, Nepal, Eastern Siberia, New Zealand and Japan report their occurrence [11]. The fungus can cause disease in a range of arthropods, for example, coleopterans, lepidoptera, hemiptera, hymenoptera, orthoptera and dipterans [91]. The latter being home to Culicidae, one of the focuses of our study.

Three species were identified for *Beauveria* fungus: *B. bassiana, B. brongniartii* and *B. pseudobassiana*. Members of *Aedes* genus, such as *Ae. aegypti* and *Ae. albopictus*, as well as those from *Culex* genus such as *Cx.*

pipiens and *Cx. quinquefasciatus*, and *Anopheles* genus such as *An. stephensi*, *An. albimanus*, *An. gambiae*, *An. arabiensis*, *An. coluzzii* and *An. funestus* are among the Culicidae fungi targets reported in the literature. *B. bassiana* stands out among the species mentioned in the adult vector control of *Ae. aegypti*, *An. stephensi* and *An. gambiae* (Table 2).

B. bassiana develops at a temperature of 22 to 26ºC, being more nutrient demanding for the growth and germination of its conidia [92]. Depending on the availability of nutrients, the infectious process on the cuticle of the host can range from 12 to 18 hours, according to Holder et al. [92].

This species was approved by the United States Environmental Protection Agency to be used as a biological tool for mosquito control [93]. The species has been extensively documented, mainly due to its adulticidal potential against Culicidae disease vectors, as some studies have shown [48, 52, 72, 94]. Due to its greater adulticidal activity, *B. bassiana* has been suggested for use in mosquitoes traps, such as ovitraps, which can also reach other life stages [95]. In adult Culicidae, infection by this fungus may reduce vector capacity, as already demonstrated for some species [23, 26], inhibit dengue virus replication within the mosquito vector through effector genes [96], increase the efficacy of chemical insecticides [27, 97], including in resistant mosquitoes [27, 28]. Despite a recognition of the fungus and an immune response by the mosquito occurring during this fungus-mosquito interaction, the vector is unable to successfully eliminate the entomopathogenic fungal infection, which causes a significant reduction in the phenoloxidase activity [94].

With respect specifically to *Ae. aegypti* vector, numerous studies have been carried out revealing the potential usefulness of the fungus against the winged mosquito phase [14, 25, 26, 43, 47, 48, 66, 94, 96, 98, 99]. From recently published studies, the use of fungus-impregnated black cloths and attractive bait in PET traps to control vector populations have been shown to efficiently reduce female survival rates under in-home conditions [47, 48]. The use of PET traps is a low-cost, easy-to-use system where the black cloth combination is based on the fact that black surfaces attract a variety of mosquito species, including *Ae. aegypti.*

Table 2. Species of the genus *Beauveria* used in the biological control of Culicidae

Species	Target Culicide	Target stage of life	Application methods	Concentrations of conidia and/or blastospores	References
B. bassiana	*An. stephensi*	Adults	Laboratory	5×10^{11} /m^2	[23]
B. bassiana	*Ae. albopictus*	Larvae and Adults	Laboratory	$1 \times 10^6 - 10^8$ /mL^{-1}	[20]
B. bassiana and *B. brongniartii*	*Ae. aegypti*	Adults	Laboratory	1×10^9 /mL	[43] and [94]
B. bassiana	*Ae. aegypti*	Adults	Laboratory	$1 \times 10^3 - 10^8$ and 2.3×10^3/mL	[96]
B. bassiana and *B. pseudobassiana*	*Ae. aegypti*	Adults	Laboratory	1×10^8 /mL	[99]
B. bassiana	*An. stephensi*	Adults	Laboratory	$1 \times 10^8 - 10^9$ /mL $^{-1}$	[52], [103] and [104]
B. bassiana	*An. stephensi*	Adults	Laboratory	$8 \times 10^8 - 2.0 \times 10^{10}$ /m^2	[51] and [106]
B. bassiana	*An. gambiae*	Adults	Field	4.6×10^{12} /m^2	[108]
B. bassiana	*An. arabiensis*	Adults	Laboratory	3.3×10^9 /cm^2	[28] and [29]
B. bassiana	*An. coluzzii*	Adults	Laboratory	1×10^8 /mL	[19] and [136]
B. bassiana	*An. stephensi* and *An. gambiae*	Larvae	Laboratory and Field	2×10^9 /cm^2	[45] and [75]
B. bassiana	*Ae. aegypti*	Adults	Laboratory and Field	2×10^9 /mL	[26] and [98]
B. bassiana	*An. gambiae*	Adults	Laboratory and Field	$2 \times 10^{10} - 7.2 \times 10^{12}$ /m^2	[57], [72] and [107]
B. bassiana	*Cx. quinquefasciatus*	Adults	Laboratory and Field	7.2×10^{12} /m^2	[4]
B. bassiana	*Ae. aegypti*	Adults	Laboratory	6×10^8 /mL^{-1}	[25]
B. bassiana	*Ae. aegypti*	Larvae	Laboratory	$1 \times 10^8 - 5 \times 10^6$ /mL^{-1}	[18], [21], [100] and [101]
B. bassiana	*Ae. aegypti*	Adults	Laboratory	1×10^{11} /m^2	[66]
B. bassiana	*Ae. aegypti*	Adults	Field	1×10^8 /mL	[47] and [48]
B. bassiana	*Cx. quinquefasciatus*	Larvae and Pupae	Laboratory	1×10^7 /mL	[5]
B. bassiana	*An. stephensi*, *Cx. quinquefasciatus* and *Ae. aegypti*	Larvae and Pupae	Laboratory	--	[102]

Table 2. (Continued)

Species	Target Culicide	Target stage of life	Application methods	Concentrations of conidia and/or blastospores	References
B. bassiana	*An. gambiae*	Adults	Laboratory	7.02 x 10^6 and 1.4 - 2.81 x 10^7/cm^2	[105]
B. bassiana	*An. stephensi*	Larvae and adults	Laboratory	1 x 10^9 and 5 x 10^8 /mL	[22] and [24]
B. bassiana	*Cx. pipiens*	Larvae	Laboratory	--	[87]
B. bassiana	*An. stephensi*	Adults	Laboratory	1 x 10^6 - 10^7 and 2 x 10^6/cm^2	[82]
B. bassiana	*An. gambiae* and *An. arabiensis*	Eggs	Laboratory	5 x 10^6 /cm^2	[84]
B. bassiana	*An. funestus*	Eggs, larvae and Adults	Laboratory	--	[73]
B. bassiana	*Ae. aegypti*	Adults	Laboratory	5 x 10^6/cm^2	[14]
B. bassiana	*Cx. pipiens*	Larvae	Laboratory and Field	1 x 10^8 - 4 x 10^8/mL	[88]
B. bassiana	*An. albimanus*	Larvae and Adults	Laboratory	2.2 x 10^7 - 4.3 x 10^8 /mL	[15]
B. bassiana	*An. arabiensis*	Adults	Field	--	[137]
B. bassiana	*An. gambiae* and *An. arabiensis*	Adults	Field	2 x 10^{10} - 8 x 10^{10}/m^2	[78]
B. bassiana	*An. gambiae*, *arabiensis* and *An. funestus*	Adults	Laboratory	--	[27]
B. bassiana	*An. gambiae*	Adults	Laboratory and Field	1 x 10^{11} - 3 x 10^{11}/m^2	[97] and [110]

B. bassiana efficiency as a self-dissemination strategy using vectors contaminated with *Ae. aegypti* is also promising. Studies have found that fungal infections acquired by this route have high mortality rates (90%) on mated females [25].

Recently, *B. bassiana* has also been studied using insect-specific scorpion neurotoxins, where *B. bassiana* recombinant lineage (Bb-AaIT) virulence was tested against *Ae. albopictus* adults [20]. In relation to the performance and effectiveness of the fungus against other cycles of this vector and *Ae. aegypti*, little has been investigated [18, 100-102]. Further studies are needed to validate the applicability of this fungus as biolarvicidal, pupicidal or ovicidal agents.

The potential usefulness of this fungus is also a promising alternative against malarial vectors, especially against winged *An. stephensi* and *An. gambiae* species [51, 52, 72, 97, 103-105]. Studies combining simple behavioral assays with electroantenograms and electropalpograms demonstrate that fungal infection may reduce the response capacity of *An. stephensi* mosquitoes in locating their hosts [103], this being correlated with a decrease in olfactory receptor neuron response [103]. This study validates previous tests which indicate this *B. bassiana* isolate is lethal to adult anopheline mosquitoes [27].

Tests that investigate the dynamics of *B. bassiana*, in *An. stephensi*, performing histopathological analysis, detect the fungal development in the legs (99%), proboscis (97%), mosquito wings (17%) and abdomen (13%), through filter paper impregnated with fungus (2.0 x $10^{10}/m^2$), adhered in a lid of Petri dishes [106]. Rhodes et al. [105] report that exposure dosage has an impact on the outcome of mosquito infections, where even in species resistant to synthetic insecticides [97, 107, 108] the fungus can be highly virulent. However, in the study conducted by Kikankie et al. [28], temperature remains an important environmental factor that probably affects both germination and fungal growth rate within resistant mosquitoes. Thus, the importance of temperature dependence on the performance of biopesticides needs to be considered in relation to the local epidemiological context.

The relatively slow killing rate of entomopathogenic fungi may be satisfactory to impact malaria transmission, since the extrinsic incubation period of the malaria parasite within the mosquito (usually 10 to 14 days in high transmission) creates a window of several days for the fungus to act [97, 109]. *B. bassiana*, activates a broad spectrum of genes, including nutrient-degrading enzymes, to explore mosquito tissue and hemolymph as nutrient sources for hyphal growth [104]. Application techniques that use spores formulated in oils (mineral or synthetic) in various mosquito resting surfaces have been shown to infect and kill most anophelines within 7-10 days of exposure [110]. Farenhorst et al. [110], describe in their study that oil formulations are considered beneficial for spore persistence in field situations, as they may protect the spores from desiccation.

In the literature, the efficacy of this entomopathogen against larvae [15, 45, 75], pupae [102] and anopheline eggs [84] has also been described, although further studies are still necessary, especially to validate its application in the field against other target stages of the most varied *Anopheles* species.

The broad spectrum of action of *B. bassiana* species in Culicidae disease vector control, becomes evident. Despite the fungus being mostly studied against *Ae. aegypti* and malaria vectors, studies have also been documented against *Cx. quinquefasciatus* and *Cx. pipiens* [5, 87, 88, 102], however these are insufficient to prove the fungus's ability to infect and kill these mosquitoes at different stages of their development.

Other Entomopathogenic Genera

Genera such as *Culicinomyces*, *Isaria*, *Lagenidium*, *Lecanicillium*, *Leptolegnia* and *Verticillium*, have been prominent in Culicidial biocontrol and show excellent toxicity in different phases of the biological cycle of these vectors. *Aspergillus* and *Fusarium* are commonly associated with agricultural pests, however these fungi can also infect and kill a variety of medically important hosts and are reported in mosquito control only at the research level (Table 3).

Verticillium genus encompasses a cosmopolitan group of ascomycetes fungi, which usually develop in moist areas with favorable temperatures. It is normally considered non-pathogenic in humans, however its health effects may be dependent on the exposure dosage and vary in individuals [111]. Its pathogenicity against Culicides has been scarcely document in the literature [111, 112], as well as the pathogenicity of *Culicinomyces* and *Lecanicillium* genera. *Culicinomyces* (Hypocreales: Cordycipitaceae), more specifically, the *C. clavisporus* species is an aquatic fungus which normally infects larvae of a variety of dipterans, including important human disease vectors such as *Ae. aegypti*, *Cx. quinquefasciatus* and

Anopheles spp. [63, 113], also being referred to as adulticide [114]. In contrast to other entomopathogenic fungi, this species is known to invade its hosts through the digestive tract after conidia ingestion [63]. Under laboratory conditions, Rodrigues et al. [63] report an ideal temperature of 25°C to support faster levels and higher conidia production values by the fungus. *Lecanicillium* genus, especially *L. muscarium* ascomycete, was described in studies by Luz et al. [115] and Leles et al. [43] as capable of reducing the survival of adult Culicides.

Studies conducted by Leles et al. [43], Blanford et al. [52], Ramirez et al. [116] and Shoukat et al. [88] are some of the more recent studies addressing the entomopathogeny of *Isaria* genus against Culicides, where the adulticidal and larvicidal action of the fungus on malaria, dengue and filariasis vectors have been evaluated. Blanford et al. [52] evaluated the virulence of 17 entomopathogenic fungi isolates, including *I. farinosus*, against the Asian mosquito *An. stephensi*. In addition, changes in food behavior after fungal infection were also evaluated. The malarial vector was exposed for six hours to the fungus by spraying (1 x 10^9 spores mL^{-1}). *I. farinosus* was the least virulent fungus against *An. stephensi* from the analyzed fungal isolates, as it had no impact on altered eating behavior. In contrast, Ramirez et al. [116] revealed the *Ae. aegypti* susceptibility to seven different *Isaria* species, with *I. javanica* and *I. amoenerosea* obtaining high pathogenicity levels [94, 116]. Data on the evaluation of fungal and chemical binary mixtures under environmental and laboratory conditions also reveal the potential usefulness of the fungus as a biolarvicide [88].

Lagenidium and *Leptolegnia* genera comprise aquatic filamentous oomycetes. Species belonging to *Leptolegnia*, such as *L. chapmanii*, infect and rapidly kill *Ae. aegypti* [117-123]. This species can be easily cultured in liquid or solid media and its zoospores, when released in larvae of this vector and kept in water at 25°C, continue infecting new larvae for more than 50 days [117]. This oomycete can infect its host both through ingestion of its zoospores and by cuticle penetration [117]. When used with temephos, results indicate a synergistic larvicidal effect against *Ae. aegypti* [119].

Table 3. Entomogenic species of genera *Aspergillus*, *Culicinomyces*, *Fusarium*, *Isaria*, *Lagenidium*, *Lecanicillium*, *Leptolegnia* e *Verticillium*, used in the biological control of Culicidae

Species	Target Culicide	Target stage of life	Application methods	Concentrations of conidia, blastospores and/or zoospores	References
A. clavatus and *flavus*	*Ae. aegypti*, *An. gambiae* and *Cx. quinquefasciatus*	Larvae and Adults	Laboratory	1.1 x 10^8 - 1.8 x 10^8 and 2.5 x 10^7- 21 x 10^7/mL	[138], [139] and [140]
A. terreus	*An. stephensi*, *Cx. quinquefasciatus* and *Ae. aegypti*	Eggs, Larvae, Pupae and Adults	Laboratory	--	[141] and [142]
A. niger	*An. stephensi*, *Ae. aegypti* and *Cx. quinquefasciatus*	Adults	Laboratory	--	[143]
C. clavisporus	*An. stephensi*, *Ae. aegypti* and *Cx. quinquefasciatus*	Eggs, Larvae and Adults	Laboratory	2 x 10^5 - 3.3 x 10^5 /mL and 5 x 10^6/cm^2	[63], [113] and [114]
F. oxysporum	*An. stephensi*, *Ae. aegypti* and *Cx. quinquefasciatus*	Larvae and Pupae	Laboratory	1 x 10^7/mL	[144] and [145]
F. pallidoroseum	*Cx. quinquefasciatus*	Adults	Laboratory	1.11 x 10^{10}/m^2	[146]
I. farinosus	*An. stephensi*	Adults	Laboratory	1 x 10^9/mL^{-1}	[52]
I. fumosorosea	*Ae. aegypti*, *Cx. pipiens* and *Cx. quinquefasciatus*	Larvae and Adults	Laboratory and Field	1 x 10^8 - 4 x 10^8 /mL	[43], [88] and [147]
Isaria spp.	*Ae. aegypti*	Adults	Laboratory	5.1 x 10^1 - 10^5/mL	[94] and [116]
L. giganteum	*Ae. aegypti*, *Culex pipiens*, *Cx. quinquefasciatus* and *An. stephensi*	Larvae and Adults	Laboratory	5.92 x 10^6/ mL	[114], [125], [126] and [127]
L. muscarium	*Ae. aegypti*, *An. arabiensis* and *Cx. quinquefasciatus*	Adults	Laboratory	--	[43] and [115]
L. chapmanii	*Ae. aegypti*	Larvae	Laboratory and Field	1.5 x 10^4 - 6.1 x 10^4 and 1.8 x 10^5 - 2.8 x 10^5/mL	[117], [118], [119], [120], [121], [122] and [123]
V. lecanii	*An. stephensi*, *Ae. aegypti* and *Cx. quinquefasciatus*	Larvae and Adults	Laboratory	--	[111] and [112]

With respect to *Lagenidium* oomycete, *L. giganteum* is among its entomopathogenic species, which has been mainly seen as a specific mosquito larvae pathogen. In more recent research, its virulence against *Cx. pipiens*, *Cx. quinquefasciatus* and *Ae. aegypti* [124-126] as well as against the anopheline *An. stephensi* has been tested [127]. Currently, *L. giganteum* was investigated using a pyrosequencing based transcriptome analysis, aiming to discover genes to improve and characterize the molecular basis of its entomopathogenicity [124]. Oomycetes express both phytopathogen canonical effector characteristics and additional virulence factors shared by invertebrate pathogens [124]. This information is promising for the development of *L. giganteum* products as an effective and environmentally sustainable tool for mosquito control. By behaving like a virulent pathogen for certain mosquito species, *L. giganteum* has been registered with the US Environmental Protection Agency in several states, including California and Florida, for use as a tool in mosquito biocontrol. It should be noted that this species was also briefly mass-produced and marketed under the name Laginex, although commercialization was preceded by several safety studies [124]. *L. giganteum*, has traditionally been mixed with other well-known entomopathogenic species due to its potential in mosquito vector biological control, such as *M. anisopliae* and *B. bassiana*. This fact is due not only to its common filamentous morphology but also to a common infectious strategy involving the destruction of the insect [124].

Challenges and Future Perspectives in the Use of Entomopathogenic Fungi

The use of entomopathogenic fungi as potential agents in Culicidae biocontrol, although widely investigated, still presents gaps and critics argue as to the longevity of fungal bioinsecticides once applied, since their slow action may be unable to provide control of these mosquitoes in different parts of the world [111]. The slow death rate associated with these bioinsecticides is still a major impediment to commercial use and large-scale application, as they may be unable to compete with faster-acting chemical insecticides [20].

Furthermore, although the entomopathogenic effects of certain fungal species are widely documented, the viability and efficacy of these in real-life conditions remain questionable [44]. Among the most significant concerns is how to minimize risks to public health with the introduction of conidia inside homes, which is still a challenge, as well as how to better distribute fungi in order to reach maximum infection rates only in target mosquitoes [44]. Therefore, formulation optimization should be critical. Moreover, a lack of knowledge surrounding the efficiency of different fungal strains under field conditions are still questionable.

Self-dissemination is an approach which may circumvent some of these questionable difficulties in the application of these fungi, where fungal dispersion and transfer would be carried out by contaminated mosquitoes. These, would spread the conidia not only in females, by mating, but in any place they made contact with, including human beings [25]. With respect to the risks some fungal entomopathogens represent for human health, researchers consider these extremely low [11, 12], since the innate immune system is the first line of defense against opportunistic infections and plays a predominant role in spore clearance in the respiratory and gastrointestinal tract, since effector cells such as basophils and mast cells reside mainly on these mucosal surfaces, where subsequent allergic reactions occur [128]. However, the appearance of symptomatology in sensitized individuals may occur and will depend on the level of exposure and other contributing factors during exposure [129].

Lastly, with respect to molecular biological technology advances, the improvement of existing tools and methods employed in disease vector controls may be perceived [41]. Gene expression analysis has shown which pathways play a role in the mosquito's antifungal response and which effector genes are linked to activation of the immune response [96]. Currently, genetic engineering has been gaining ground in entomopathogenic fungal studies, where these have been genetically modified to increase their virulence against mosquitoes [20].

We conclude that *Metarhizium* and *Beauveria* are the best studied genera over the last 11 years reported in this review. *M. anisopliae* species is favorable for use as a biolarvicide and adulticide, especially against *Ae. aegypti* as well as *An. stephensi* and *An. gambiae* malarial vectors. For these Culicidae, *B. bassiana* has also been shown to be strongly virulent, however with adulticide applicability. From the remaining entomopathogens, the oomycete *L. chapmanii*, appears as a promising candidate in the infectivity and death of *Ae. aegypti* larvae. In spite of the potential effectiveness of these entomopathogens and the many methodological methods developed and tested, including simulated intra-domiciliary conditions, their application in substitution of chemical products in the current control programs, is still questionable, especially, in relation to their public health safety.

REFERENCES

[1] Corbel, V., Fonseca, D. M., Weetman, D., Pinto, J., Achee, N. L., Chandre, F., Coulibaly, M. B., Dusfour, I., Grieco, J., Juntarajumnong, W., Lenhart, A., Martins, A. J., Moyes, C., Ng Ching, L., Pinto, J., Raghavendra, K., Vatandoost, H., Vontas, J., Weetman, D., Fouque, F., Velayudhan, R., David, J. P. Tracking Insecticide Resistance in Mosquito Vectors of Arboviruses: The Worldwide Insecticide resistance Network (WIN). *PLoS Negl. Trop. Dis.* 2016, *10*, p. 1-4, doi:10.1371/journal.pntd.0005054.

[2] Corbel, V., Fonseca, D. M., Weetman, D., Pinto, J., Achee, N. L., Chandre, F., Coulibaly, M. B., Dusfour, I., Grieco, J., Juntarajumnong, W., Lenhart, A., Martins, A. J., Moyes, C., Ng Ching, L., Raghavendra, K., Vatandoost, H., Vontas, J., Muller, P., Kasai, S., Fouque, F., Velayudhan, R., Durot, C., David, J. P. International workshop on insecticide resistance in vectors of arboviruses, December 2016, Rio de Janeiro, Brazil. *Parasit. Vectors* 2017, *10,* p. 2-16, 10.1186/s13071-017-2224-3.

[3] Consoli, R. A. G. B., Oliveira, R. L. *Principais mosquitos de importância sanitária no Brasil.* 1ª Ed., Fiocruz: Rio de Janeiro, Brasil, 1998, pp. 228. [*Main mosquitoes of sanitary importance in Brazil.*]

[4] Howard, A. F. V., Guessan, R. N., Koenraad, C. J. M. Asidi, A., Farenhorst, M., Akogbéto, M., Thomas, M. B., Knols, B. G. J., Takken, W. The entomopathogenic fungus *Beauveria bassiana* reduces instantaneous blood feeding in wild multi-insecticide-resistant *Culex quinquefasciatus* mosquitoes in Benin, West Africa. *Parasit. Vectors* 2010, *3*, p- 2-11, doi.org/10.1186/1756-3305-3-87.

[5] Vivekanandhan, P., Kavitha, T., Karthi, S., Senthil-Nathan, S., Shivakumar, M. S. Toxicity of *Beauveria bassiana*-28 Mycelial Extracts on Larvae of *Culex quinquefasciatus* Mosquito (Diptera: Culicidae). *Int. J. Environ. Res. Public. Health* 2018, *15*, p. 2-11, doi:10.3390/ijerph15030440.

[6] Barros, N. M., Vargas, L. R. B., Schrank, A., Boldo, J. T., Specht, A. Fungos como agentes de controle de pragas. In: *Fungos: uma introdução à biologia, bioquímica e biotecnologia*, 2ª ed., Esposito, E., Azevedo, J. L., Educs: Caxias do Sul, Brasil, 2010, pp. 491-532. [Fungi as pest control agents. In: *Fungi: an introduction to biology, biochemistry and biotechnology*]

[7] Gao, L. A. Novel Method to Optimize Culture Conditions for Biomass and Sporulation of the Entomopathogenic Fungus *Beauveria bassiana* IBC1201. *Braz. J. Microbiol.* 2011, *42*, p. 1574-1584.

[8] Rodríguez-Pérez, M. A., Howard, A. F. V., Reyes-Villanueva, F. Controle Biológico de Vetores de Dengue. Em: *Manejo Integrado de Pragas e Controle de Pragas - Táticas Atuais e Futuras.* Larramendy, M. L., Soloneski, S. Rijeka, Croácia, 2012, pp. 241-270.

[9] Greenfield, B. P. J., Peace, A., Evans, H., Dudley, E. d., Ansari, M. A., Butt, T. M. Identification of *Metarhizium* strains highly efficacious against *Aedes*, *Anopheles* and *Culex* larvae, *Biocontrol*

Sci. and Technol. 2015, *25*, p. 487-502, doi.org/10.1080/09583157.2014.989813.

[10] Alkhaibari, A. M., Carolino, A. T., Bull, J. C., Samuels, R. I., Butt, T. M. Differential Pathogenicity of *Metarhizium* Blastospores and Conidia Against Larvae of Three Mosquito Species. *J. Med. Entomol.* 2017, *54*, p. 696-704, doi: 10.1093/jme/tjw223.

[11] Zimmermann, G. Review on safety of the entomopathogenic fungi *Beauveria bassiana* and *Beauveria brongniartii*. *Biocontrol. Sci. Techn.* 2007, *17*, 553-596.

[12] Darbro, J. M., Thomas, M. B. Spore Persistence and Likelihood of Aeroallergenicity of Entomopathogenic Fungi Used for Mosquito Control. *Am. J. Trop. Med. Hyg.* 2009, *80*, p. 992–997.

[13] Mora, M. A. E., Castilho, A. M. C., Fraga, M. E. Classification and infection mechanism of entomopathogenic fungi. *Arq. Inst. Biol.* 2017, *84*, p.1-10, doi: 10.1590/1808-1657000552015.

[14] Paula, A. R., Brito, E. S., Pereira, C. R., Carrera, M. P., Samuels, R. I. Susceptibility of adult *Aedes aegypti* (Diptera: Culicidae) to infection by *Metarhizium anisopliae* and *Beauveria bassiana*: prospects for Dengue vector control. *Biocontrol Sci. Techn.* 2008, *18*, p. 1017-1025, doi:10.1080/09583150802509199.

[15] Vázquez-Martínez, M. G., Rodríguez-Meneses, A., Rodríguez, A. D. Rodríguez, M. H. Lethal effects of *Gliocladium virens*, *Beauveria bassiana* and *Metarhizium anisopliae* on the malaria vector *Anopheles albimanus* (Diptera: Culicidae). *Biocontrol Sci. Techn.* 2013, *23*, p. 1098-1109, doi.org/10.1080/09583157.2013.822470.

[16] Abdou, M. A., Salem, D. A., Abdallah, F. I., Diwan, N. M. L. The Biochemical Effects of *Beauveria bassiana* and *Metarhizium anisopliae* on 3rd instar Larvae of *Culex pipiens* L. (Diptera: Culicidae). *Egypt. Acad. J. Biolog. Sci.* 2017, *10*, p. 35-44.

[17] López-Lastra, C. C., Scorsetti, A. C. Revisión de los hongos entomophthorales (Zygomycota: Zygomycetes) patógenos de insectos de la República Argentina. *Bol. Soc. Argent. Bot.* 2007, *42*, p. 33-37. [Review of insect pathogenic entomophthoral fungi (Zygomycota: Zygomycetes) from the Argentine Republic.]

[18] Banu, A. N., Balasubramanian, C. Myco-synthesis of silver nanoparticles using *Beauveria bassiana* against dengue vector, *Aedes aegypti* (Diptera: Culicidae). *Parasitol Res.* 2014, *113*, p. 2869–2877, doi:10.1007/s00436-014-3948-z.

[19] Valero-Jiménez, C. A., Faino, L., Veld, D. S. I., Smit, S., Zwaan, B. J., Kan, J. A. L. V. Comparative genomics of *Beauveria bassiana*: uncovering signatures of virulence against mosquitoes. *BMC Genomics* 2016, *17*, p. 2-11, doi:10.1186/s12864-016-3339-1.

[20] Deng, S Q., Cai, Q. D., Deng, M. Z., Huang, Q., Peng, H. J. Scorpion neurotoxin AaIT-expressing *Beauveria bassiana* enhances the virulence against *Aedes albopictus* mosquitoes. *AMB Express* 2017, *7*, p. 2-10, doi:10.1186/s13568-017-0422-1.

[21] Daniel, J. F. C., Silva, A. A., Nagakawa, D. H., Medeiros, L. S., Carvalho, M. G., Tavares, L. J., Abreu, L. M., Rodrigues-Filho, E. Larvicidal Activity of *Beauveria bassiana* Extracts against *Aedes aegypti* and Identification of Beauvericins. *J. Braz. Chem. Soc.* 2017, *28*, p. 1003-1013, doi.org/10.21577/0103-5053.20160253.

[22] Veys-Behbahani, R., Sharififard, M., Dinparast-Djadid, N., Shamsi, J., Fakoorziba, M. R. Laboratory evolution of the entomopathogenic fungus Beauveria bassiana against Anopheles stephensi larvae (Diptera: Culicidae). *Asian Pac. J Trop. Dis.* 2014, *4*, p. S799-S802.

[23] Blanford, S., Shi, W., Christian, R., Marden, J. H., Koekemoer, L. L., Brooke, B. D., Coetzee, M., Read, A. F., Thomas, M. B. Lethal and Pre-Lethal Effects of a Fungal Biopesticide Contribute to Substantial and Rapid Control of Malaria Vectors. *Plos One*, 2011, *6*, p.1-11, doi:10.1371/journal.pone.0023591.

[24] Vogels, C. B. F., Bukhari, T., Koenraadt, C. J. M. Fitness consequences of larval exposure to *Beauveria bassiana* on adults of the malaria vector *Anopheles stephensi*. *J. Invertebr. Pathol.* 2014, *119*, p. 19-24, doi.org/10.1016/j.jip.2014.03.003.

[25] García-Munguía, A. M., Garza-Hernández, C. A., Rebollar-Tellez, E. A., Rodríguez-Pérez, M. A., Reyes-Villanueva, F. Transmission of *Beauveria bassiana* from male to female *Aedes aegypti*

mosquitoes. *Parasit. Vectors* 2011, *4*, p. 2-6, doi.org/10.1186/1756-3305-4-24.

[26] Darbro, J. M., Johnson, P. H., Thomas, M. B., Ritchie, S. A., Kay, B. H., Ryan, P. A. Effects of *Beauveria bassiana* on Survival, Blood-Feeding Success, and Fecundity of *Aedes aegypti* in Laboratory and Semi-Field Conditions. *Am. J. Trop. Med. Hyg.* 2012, *86*, p. 656 -664, doi:10.4269/ajtmh.2012.11-0455.

[27] Farenhorst, M., Mouatcho, J. C., Kikankie, C. K., Brooke, B. D., Hunt, R. H., Thomas, M. B., Koekemoer, L. L., Knols, B. G. J., Coatzee, M. Fungal Infection Counters Insecticide Resistance in African Malaria Mosquitoes. *Proc. Natl. Acad. Sci. USA* 2009, *106*, p. 17443-17447, doi.org/10.1073/pnas.0908530106.

[28] Kikankie, C. K., Brooke, B. D., Knols, B. G. J., Koekemoer, L. L., Farenhorst, M., Hunt, R. H., Thomas, M. B., Coetzee, M. The infectivity of the entomopathogenic fungus *Beauveria bassiana* to insecticide-resistant and susceptible *Anopheles arabiensis* mosquitoes at two different temperatures. *Malar. J.* 2010, *9*, p. 2-9.

[29] Nardini, L., Blanford, S., Coetzee, M., Koekemoer, L. L. Effect of *Beauveria bassiana* infection on detoxification enzyme transcription in pyrethroid resistant *Anopheles arabiensis*: a preliminary study. *Soc. Trop. Med. Hyg.* 2014, p. 1-7, doi:10.1093/trstmh/tru021.

[30] Boucias, D. G., Pendland, J. C. *Princípios da Patologia de Insetos*. Massachusetts: Kluwer Academic Publishers, 1998, pp. 551.

[31] Zimmermann, G. Review on safety of the entomopathogenic fungus *Metarhizium anisopliae*. *Biocontrol Sci. Technol.* 2007, *17*, p. 879-920, doi.org/10.1080/09583150701593963.

[32] Bilal, H., Hassan, S. A., Khan, I. A. Isolation and efficacy of entomopathogenic fungus (*Metarhizium anisopliae*) for the control of Aedes albopictus Skuse larvae: suspected dengue vector in Pakistan. *Asian Pac J Trop Biomed* 2012, *2*, p. 298-300, doi:10. 1016/S2221-1691(12)60026-4.

[33] Garza-Hernández, J. A., Rodríguez-Pérez, M. A., Salazar, M. I., Russell, T. L., Adeleke, M. A., Luna-Santillana, E. J., Reyes-Villanueva, F. Vectorial Capacity of *Aedes aegypti* for Dengue

Virus Type 2 Is Reduced with Co-infection of *Metarhizium anisopliae*. *Plos Negl. Trop. Dis.* 2013, *7*, p. 1-5, doi:10.1371/journal.pntd.0002013.

[34] Alkhaibari, A. M., Lordc, A. M., Maffeisc, T., Bulla, J. C., Olivaresd, F. L., Samuels, R. I., Butt, T. Highly specific host-pathogen interactions influence *Metarhizium brunneum* blastospore virulence against *Culex quinquefasciatus* larvae. *Virulence* 2018, *9*, p. 1449-1467, doi.org/10.1080/21505594.2018.1509665.

[35] Vega, F. E., Goettel, M. S., Blackwell, M., Chandler, D. Entomopatógenos fúngicos: novos insights sobre sua ecologia. *Fungal Ecol* 2009, *2*, p. 149–159. [Fungal entomopathogens: new insights into their ecology.]

[36] Kannan, S. K., Murugan, K., Kumar, A. N., Ramasubramanian, N., Mathiyazhagan, P. Adulticidal effect of fungal pathogen, *Metarhizium anisopliae* on malarial vector *Anopheles stephensi* (Diptera: Culicidae). *Afr. J. of Biotechnol.* 2008, *7*, p. 838-841.

[37] Paula, A. R., Carolino, A. T., Paula, C. O., Samuels, R. I. The combination of the entomopathogenic fungus *Metarhizium anisopliae* with the insecticide Imidacloprid increases virulence against the dengue vector *Aedes aegypti* (Diptera: Culicidae). *Parasit. Vectors* 2011, *4*, p. 1-8, doi.org/10.1186/1756-3305-4-8.

[38] Falvo, M. L., Pereira-Junior, R. A., Rodrigues, J., Lopez Lastra, C.C., Garcia, J. J., Fernandes, É. K. K., Luz, C. UV-B radiation reduces in vitro germination of *Metarhizium anisopliae* s.l. but does not affect virulence in fungus treated *Aedes aegypti* adults and development on dead mosquitoes. *J. Appl. Microbiol.* 2016, *121*, p. 1710-1717.

[39] Scholte, E. J., Knols, B. G. J., Samson, R. A., Takken, W. Entomopathogenic fungi for mosquito control: A review. *J. Insect Sci.* 2004, 4, pp. 24.

[40] Roy, H. E., Steinkraus, D. C., Eilenberg, J., Hajek, A. E., Pell, J. K. Bizarre interactions and endgames: entomopathogenic fungi and their arthropod hosts. *Annu. Rev. Entomol.* 2006, *51*, p. 331-57, doi:10.1146/annurev.ento.51.110104.150941.

[41] Huang Yan-Jang, S., Higgs, S., Vanlandingham, DL. Biological Control Strategies for Mosquito Vectors of Arboviruses. *Insects* 2017, *8*, p. 2-25, doi:10.3390/insects8010021.

[42] Sbaraini, N., Muniz Guedes, R. L., Andreis, F. C., Junges, A. Morais, G. L., Vainstein, M. H., Vasconcelos, A. T. R., Schrank, A. Secondary metabolite gene clusters in the entomopathogen fungus *Metarhizium anisopliae*: genome identification and patterns of expression in a cuticle infection model. *BMC Genomics* 2016, *17*, p. 400-462, doi:10.1186/s12864-016-3067-6.

[43] Leles, R. N., Sousa, N. A., Rocha, L. F. N., Santos, A. H., Silva, H. H. G., Luz, C. Pathogenicity of some hypocrealean fungi to adult *Aedes aegypti* (Diptera: Culicidae). *Parasitol Res.* 2010, 107, p. 1271-1274, doi:10.1007/s00436-010-1991-y.

[44] Lwetoijera, D. W., Sumaye, R. D., Madumla, E. P., Kavishe, D. R. Mnyone, L. L., Russell, T. L., Okumu, F. O. An extra-domiciliary method of delivering entomopathogenic fungus, *Metharizium anisopliae* IP 46 for controlling adult populations of the malaria vector, *Anopheles arabiensis*. *Parasit. Vectors* 2010, *3*, p. 2-6, doi.org/10.1186/1756-3305-3-18.

[45] Bukhari, T., Takken, W., Koenraadt, C. J. M. Development of *Metarhizium anisopliae* and *Beauveria bassiana* formulations for control of malaria mosquito larvae. *Parasit. Vectors* 2011, *4*, p. 1-14.

[46] Paula, A. R., Carolino, A. T., Silva, C. P., Pereira, C. R., Samuels, R. I. Testing fungus impregnated cloths for the control of adult *Aedes aegypti* under natural conditions. *Parasit. Vectors* 2013, *6*, p. 1-6.

[47] Silva, L. E. I., Paula, A. R., Ribeiro, A., Butt, T. M., Silva, C. P. Samuels RI. A new method of deploying entomopathogenic fungi to control adult *Aedes aegypti* mosquitoes. *J. Appl. Entomol.* 2017, p. 1-8, doi: 10.1111/jen.12402.

[48] Paula, A. R., Silva, L. E. I., Reibeiro, A., Butt, T. M., Silva, C. P., Samuels, R. I. Improving the delivery and efficiency of fungus-impregnated cloths for control of adult *Aedes aegypti* using a

synthetic attractive lure. *Parasit. Vectors* 2018, *11*, p. 2-9, doi.org/10.1186/s13071-018-2871-z.

[49] Luz, C., Tai, M. H. H., Santos, A. H., Silva, H. H. G. Impact of moisture on survival of *Aedes aegypti* eggs and ovicidal activity of *Metarhizium anisopliae* under laboratory conditions. *Mem. Inst. Oswaldo Cruz* 2008, *103*, p. 214-215.

[50] Carolino, A. T., Paula, A. R., Silva, C. P., Butt, T. M., Samuels, R. Monitoring persistence of the entomopathogenic fungus *Metarhizium anisopliae* under simulated field conditions with the aim of controlling adult *Aedes aegypti* (Diptera: Culicidae). *Parasit. Vectors* 2014, *7*, p. 2-7, doi.org/10.1186/1756-3305-7-198.

[51] Heinig, R. L., Thomas, M. B. Interactions between a fungal entomopathogen and malaria parasites within a mosquito vector. *Malar. J.* 2015, *14*, p. 2-10, doi:10.1186/s12936-014-0526-x.

[52] Blanford, S., Jenkins, N. E., Read, A. F., Thomas, M. B. Evaluating the lethal and pre-lethal effects of arange of fungi against adult *Anopheles stephensi* mosquitoes. *Malar. J.* 2012, *11*, p. 1-10, doi.org/10.1186/1475-2875-11-365.

[53] Wang, Y. H., Hu, Y., Xing, L-S., Jiang, H., Hu, S. N., Raikhel, A. S., Zou, Z. A Critical Role for CLSP2 in the Modulation of Antifungal Immune Response in Mosquitoes. *Plos Pathog*. 2015, *11*, p. 2-20, doi:10.1371/journal.ppat.1004931.

[54] Wang, Z., Jin, K., Xia, Y. Transcriptional analysis of the conidiation pattern shift of the entomopathogenic fungus *Metarhizium acridum* in response to different nutrientes. *BMC Genomics* 2016, *17*, p. 2-11, doi:10.1186/s12864-016-2971-0.

[55] Reyes-Villanueva, F., Garza-Hernandez, J. A., Garcia-Munguia, A. M., Tamez-Guerra, P., Howard, A. F. V., Rodriguez-Perez, M. A. Dissemination of *Metarhizium anisopliae* of low and high virulence by mating behavior in *Aedes aegypti. Parasit. Vectors* 2011, *4*, p. 2-7.

[56] Garza-Hernández, J. A., Reyes-Villanueva, F., Russell, T. L., Braks, M. A. H., Garcia-Munguia, A. M., Rodríguez-Pérez, M. A. Copulation Activity, Sperm Production and Conidia Transfer in

Aedes aegypti Males Contaminated by *Metarhizium anisopliae*: A Biological Control Prospect. *Plos Negl. Trop. Dis.* 2015, *9*, p. 1-14, doi:10.1371/journal.pntd.0004144.

[57] Mnyone, L. L., Kirby, M. J., Lwetoijera, D. W., Mpingwa, M. W., Simfukwe, E. T., Knols, B. G. J., Takken, W., Russell, T. L. Tools for delivering entomopathogenic fungi to malaria mosquitoes: effects of delivery surfaces on fungal efficacy and persistence. *Malar. J.* 2010, *9*, p. 1-7, doi.org/10.1186/1475-2875-9-246.

[58] Bilgo, E., Lovett, B., Leger, R. J. St., Sanon, A., Dabiré, R. K. Diabaté, A. Native entomopathogenic *Metarhizium* spp. from Burkina Faso and their virulence against the malaria vector *Anopheles coluzzii* and non-target insects. *Parasit. Vectors* 2018, *11*, p. 2-6, doi.org/10.1186/s13071-018-2796-6.

[59] Kamalakannan S., Murugan, K. Laboratório e avaliação de campo de *Metarhizium anisopliae* para o controle do vetor da dengue, *Aedes aegypti* (Insecta: Diptera: Culicidae) *Toxicol. Environ. Chem.* 2011, *93*, p. 1195-1201, doi: 10.1080/02772248.2011.580577.

[60] Lee, S. J., Kim, S., Yu, J. S., Kim, J. C., Nai, Y. S., Kim, J. S. Biological control of Asian tiger mosquito, *Aedes albopictus* (Diptera: Culicidae) using *Metarhizium anisopliae* JEF-003 millet grain. *J Asia Pac Entomol.* 2015, *18*, p. 217-221.

[61] Butt, T. M., Greenfield, B. P. J., Greig, C., Maffeis, T. G. G., Taylor, J. W. D., Piasecka, J., Dudley, Ed., Abdulla, A., Dubovskiy, I. M., Garrido-Jurado, I., Quesada-Moraga, E., Penny, M. W., Eastwood, D. *Metarhizium anisopliae* Pathogenesis of Mosquito Larvae: A Verdict of Accidental Death. *Plos One* 2013, *8*, p. 1-11, doi:10.1371/journal.pone.0081686.

[62] Greenfield, B. P. J., Lord, A. M., Dudley, E. d., Butt, T. M. Conidia of the insect pathogenic fungus, *Metarhizium anisopliae*, fail to adhere to mosquito larval cuticle. *R. Soc. open sci.* 2014, *1*, p. 2-9, doi.org/10.1098/rsos.140193.

[63] Rodrigues, J., Luz, C., Humberb, R. A. New insights into the in vitro development and virulence of *Culicinomyces* spp. as fungal pathogens of *Aedes aegypti*. *J. Invertebr. Pathol.* 2017,*146*, p. 7-13.

[64] Alkhaibari, A. M., Carolino, A. T., Yavasoglu, S. I., Maffeis, T. Mattoso, T. C., Bull, J. C., Samuels, R. I., Butt, T. M. *Metarhizium brunneum* Blastospore Pathogenesis in *Aedes aegypti* Larvae: Attack on Several Fronts Accelerates Mortality. *Plos. Pathog.* 2016, *12*, p. 1-19, doi:10.1371/journal.ppat.1005715.

[65] Gomes, S. A., Paula, A. R., Ribeiro, A., Moraes, C. O. P., Santos, J. W. A. B., Silva, C. P., Samuels, R. I. Neem oil increases the efficiency of the entomopathogenic fungus *Metarhizium anisopliae* for the control of *Aedes aegypti* (Diptera: Culicidae) larvae. *Parasit. Vectors* 2015, 8, p. 2-8, doi:10.1186/s13071-015-1280-9.

[66] Darbro, J. M., Graham, R. Kay, B. H. Ryan, P. A., Thomas, M. B. Evaluation of entomopathogenic fungi as potential biological control agents of the dengue mosquito, *Aedes aegypti* (Diptera: Culicidae). *Biocontrol Sci. Technol.* 2011, *21*, p. 1027-1047, doi: 10.1080/0958 3157.2011.597913.

[67] Luz, C., Tai, M. H. H., Santos, A. H., Rocha, L. F. N., Albernaz, D. A. S., Silva, H. H. G. Ovicidal activity of entomopathogenic Hyphomycetes on *Aedes aegypti* (L.) (Diptera: Culicidae) under laboratory conditions. *J. Med. Entomol.* 2007, *44*, p. 799-804.

[68] Albernaz, D. A. S., Tai, M. H. H., Luz, C. Enhanced ovicidal activity of an oil formulation of the fungus *Metarhizium anisopliae* on the mosquito *Aedes aegypti. Med Vet Entomol.* 2009, *23*, p. 141-147, doi: 10.1111/j.1365-2915.2008.00792.x.

[69] Santos, A. H., Tai, M. H. H., Rocha, L. F. N., Silva, H.H.G., Luz, C. Dependence of *Metarhizium anisopliae* on high humidity for ovicidal activity on *Aedes aegypti. Biol. Control* 2009, *59*, p.37-42.

[70] Leles, R. N., D' Alessandro, W. B., Luz, C. Effects of *Metarhizium anisopliae* conidia mixed with soil against the eggs of *Aedes aegypti. Parasitol Res.* 2012, *110*, p. 1579-1582, doi: 10.1007/s00436-011-2666-z.

[71] Sousa, N. A., Lobo, L. S., Rodrigues, J. Luz, C. New insights on the effectiveness of *Metarhizium anisopliae* formulation and application against *Aedes aegypti* eggs. *Lett. Appl. Microbiol.* 2013, *57*, p. 193-199.

[72] Mnyone, L. L., Kirby, M. J., Lwetoijera, D. W., Mpingwa, M. W., Knols, B. G. J., Takken, W., Russell, T. L. Infection of the malaria mosquito, *Anopheles gambiae*, with two species of entomopathogenic fungi: effects of concentration, co-formulation, exposure time and persistence. *Malar. J.* 2009, *8*, p. 1-12, doi:10. 1186/1475-2875-8-309.

[73] Mouatcho, J. C., Koekemoer, L. L., Coetzee, M., Brooke, B. D. The effect of entomopathogenic fungus infection on female fecundity of the major malaria vector, *Anopheles funestus*. *Afr. Entomol.* 2011 *19*, p.725-729, doi.org/10.4001/003.019.0311.

[74] Yasuoka, J. Jimba, M. Levins, R. Application of loop analysis for evaluation of malaria control interventions. *Malar. J.* 2014, *13*, p. 2-15, doi.org/10.1186/1475-2875-13-140.

[75] Bukharil, T., Middelman, A., Koenraadt, C. J. M. Takken, W., Knols, B. G. J. Factors affecting fungus-induced larval mortality in *Anopheles gambiae* and *Anophelesstephensi*. *Malar. J.* 2010, *9*, p. 2-15, doi.org/10.1186/1475-2875-9-22.

[76] Thomas, M. B., Read, A. F. Can fungal biopesticides control malária. *Nature* 2007, *5*, p. 377-383.

[77] Farenhorst, M., Farina, D., Scholte, E., Takken, W., Hunt, R. H., Coetzee, M., Knols, B. G. J. African Water Storage Pots for the Delivery of the Entomopathogenic Fungus Metarhizium anisopliae to the Malaria Vectors *Anopheles gambiae* s.s. and *Anopheles funestus*. *Am. J. Trop. Med. Hyg.* 2008, *78*, p. 910-916.

[78] Mnyone, L. L., Lyimo, I. N., Lwetoijera, D. W., Mpingwa, M. W., Nchimbi, N., Hancock, P. P., Russell, T. L., Kirby, M. J. Takken, W., Koenraadt, C. J. M. Exploiting the behaviour of wild malária vectors to achieve high infection with fungal biocontrol agents. *Malar. J.* 2012, *11*, p. 2-11.

[79] Mmbando, A. S., Okumu, F. O., Mgando, J. P., Sumaye, R. D., Matowo, N. S., Madumla, E., Kaindoa, E., Kiware, S. S., Lwetoijera, D. W. Effects of a new outdoor mosquito control device, the mosquito landing box, on densities and survival of the malária

vector, *Anopheles arabiensis*, inside controlled semi-field settings. *Malar. J.* 2015, *14*, p. 2-13, doi:10.1186/s12936-015-1013-8.

[80] Harris, C., Lwetoijera, D., Dongus, S., Matowo, N. S., Lorenz, L. M. Devine, G. J., Majambere, S. Sterilising effects of pyriproxyfen on *Anopheles arabiensis* and its potential use in malaria control. *Parasite Vector* 2013, *6*, p. 2-8.

[81] Lwetoijera, D., Harris, C., Kiware, S. Dongus, S., Devine, G. J., McCall, P. J., Majambere, S. Effective autodissemination of pyriproxyfen to breeding sites by the exophilic malaria vector *Anopheles arabiensis* in semi-field settings in Tanzania. *Malar J.* 2014, *13*, p. 2-10.

[82] Bell, A. S., Blanford, S., Jenkins, N., Thomas, M. B., Read, A. F. Real-time quantitative PCR for analysis of candidate fungal biopesticides against malaria: Technique validation and first applications. *J. Invertebr. Pathol.* 2009, 100, p. 160-168, doi:10. 1016/j.jip.2009.01.006.

[83] Bilgo, E., Lovett, B., Bayili, K., Millogo, A. S., Saré, I., Dabiré, R. K., Sanon, A., Leger, R. J. St. Diabate, A. Transgenic *Metarhizium pingshaense* synergistically ameliorates pyrethroid-resistance in wild-caught, malaria-vector mosquitoes. *PLoS One* 2018, *13*, doi.org/10.1371/journal.pone.0203529.

[84] Luz, C., Mnyone, L. L., Russell, T. L. Survival of anopheline eggs and their susceptibility to infection with *Metarhizium anisopliae* and *Beauveria bassiana* under laboratory conditions. *Parasitol. Res.* 2011, *109*, p. 751-758, doi:10.1007/s00436-011-2318-3.

[85] Murugan, K., Kovendan, K., Vincent, S., Barnard, D. R. Biolarvicidal and pupicidal activity of *Acalypha alnifolia* Klein ex Willd. (Family: Euphorbiaceae) leaf extract and microbial insecticide, *Metarhizium anisopliae* (Metsch.) against malaria fever mosquito, *Anopheles stephensi* Liston. (Diptera: Culicidae). *Parasitol. Res.* 2012, *110*, p. 2263-2270, doi:10.1007/s00436-011-2758-9.

[86] Mohanty, S. S., Raghavendra, K., Dash, A. P. Induction of chymoelastase (Pr1) of *Metarhizium anisopliae* and its role in

causing mortality to mosquito larvae. *World J. Microbiol. Biotechnol.* 2008, *24*, p. 2283-2288, 10.1007/s11274-008-9742-2.

[87] Abdou, M. A., Salem, D. A., Abdallah, F. I., Diwan, N. M. L. The Biochemical Effects of *Beauveria bassiana* and *Metarhizium anisopliae* on 3rd instar Larvae of *Culex pipiens* L. (Diptera: Culicidae). *Egypt. Acad. J. Biolog. Sci.* 2017, *10*, p. 35-44.

[88] Shoukat, R. F., Freed, S., Ahmad, K. W., Rehman, A. Assessment of Binary Mixtures of Entomopathogenic Fungi and Chemical Insecticides on Biological Parameters of *Culex pipiens* (Diptera: Culicidae) under Laboratory and Field Conditions. *J. Zool.* 2018, *50*, p. 299-309, doi.org/10.17582/journal.pjz/2018.50.1.299.309.

[89] Barra-Bucarei, L. Vergara, P., Cortes, A. Conditions to optimize mass production of *Metarhizium anisopliae* (Metschn.) Sorokin 1883 in different substrates. *Chil. j. agric. res.* 2016, *76*, p. 448-454, doi:10.4067/S0718-58392016000400008.

[90] Soares, F. B., Monteiro, A. C., Barbosa, J. C., Mochi, D. A. Population density of *Beauveria bassiana* in soil under the action of fungicides and native microbial populations. *Acta Sci. Agron.* 2017, *39*, p. 465-474, doi:10.4025/actasciagron.v39i4.32816.

[91] Safavi, S. A., Shah, F. A., Pakdel, A. K., Rasoulian, G. R., Bandani, A. R., Butt, T. M. Efect of nutrition on growth and virulence of the entomopathogenic fungus *Beauveria bassiana. FEMS Microbiol.* 2007, *270*, p. 116-123, doi:10.1111/j.1574-6968.2007.00666.x.

[92] Holder, D. J., Kirkland, B. H., Lewis, M. W., Keyhani, N. O. Surface characteristics of the entomopathogenic fungus *Beauveria* (Cordyceps) *bassiana*. Microbiology 2007, *153*, p. 3448–3457, doi:10.1099/mic.0.2007/008524-0.

[93] Singh, R. K., Dhama, K., Khandia, R., Munjal, A., Karthik, K., Tiwari, R., Chakraborty, S., Malik, Y. S., Bueno-Marí, R. Prevention and Control Strategies to Counter ZikaVirus, a Special Focus on Intervention Approaches against Vector Mosquitoes-Current Updates. *Front. Microbiol.* 2018, *9*, p. 1-22, doi: 10.3389/fmicb.2018.00087.

[94] Ramirez, J. L., Dunlap, C. A., Muturi, E. J., Barletta, A. B. F., Rooney, A. P. Entomopathogenic fungal infection leads to temporospatial modulation of the mosquito immune system. *Plos Negl. Trop. Dis.* 2018, *12*, p.1-24, doi.org/10.1371/journal.pntd. 0006433.

[95] Snetselaar, J., Andriessen, R., Suer, R. A., Osinga, A. J. Knols, B. G. J. Farenhorst, M. Development and evaluation of a novel contamination device that targets multiple life-stages of *Aedes aegypti. Parasit. Vectors* 2014, *7*, p. 2-10, doi.org/10.1186/1756-3305-7-200.

[96] Dong, Y., Morton Junior, J. C., Ramirez, J. L., Souza-Neto, J. A., Dimopoulos, G. The entomopathogenic fungus *Beauveria bassiana* activate Toll and JAK-STAT pathway controlled effector genes and antidengue activity in *Aedes aegypti. Insect. Biochem. Mol. Biol.* 2012, *42*, p. 126-132.

[97] Farenhorst, M., Knols, B. G. J., Thomas, M. B., Howard, A. F. V., Takken, W., Rowland, M., N'Guessan, R. Synergy in Efficacy of Fungal Entomopathogens and Permethrin against West African Insecticide-Resistant *Anopheles gambiae* Mosquitoes. *PLoS ONE* 2010, *5*, p. 1-10, doi:10.1371/journal.pone.0012081.

[98] Wang, C., Wang, S. Insect Pathogenic Fungi: Genomics, Molecular Interactions, and Genetic Improvements. *Annu. Rev. Entomol.* 2017, *62*, p.73-90, doi.org/10.1146/annurev-ento-031616-035509.

[99] Tejeda-Reyes, M. A., Rodríguez-Maciel, J. C., Alatorre-Rosas, R., Lagunes-Tejeda, A., Vargas-Hernández, M., Silva-Aguayo, G. I. A new methodology to evaluate entomopathogenic fungi and formulated insecticides to control adults of *Aedes aegypti* (Diptera: Culicidae). *Fla. Entomol.* 2018, *101*, p. 511-514, doi.org/10.1653/ 024.101.0311.

[100] Gimenes, D. C., Alexandrino, T. D., Machado, A. C., Varéa, G. S. Induction of Proteases from Beauveria bassiana by *Aedes aegypti* larvae and cuticle cicadas. *Biochemi. Biotechnol. Reports* 2014, *3*, p. 41-47, doi:10.5433/2316-5200.2014v3n1p41.

[101] Pereira, C. R., Paula, A. R., Gomes, S. A., Pedra Jr., P. C. O, Samuels, R. I. The potential of *Metarhizium anisopliae* and *Beauveria bassiana* isolates for the control of *Aedes aegypti* (Diptera: Culicidae) larvae. *Biocontrol Science and Technology* 2009, *19*, p. 881-886, doi:10.1080/09583150903147659.

[102] Ragavendran, C., Dubeyb, N. K., Natarajan, D. *Beauveria bassiana* (Clavicipitaceae): a potente fungal agent for controlling mosquito vectors of *Anopheles stephensi, Culex quinquefasciatus* and *Aedes aegypti* (Diptera: Culicidae). *RSC Adv*. 2017, *7*, p. 3838-3851, doi: 10.1039/c6ra25859j.

[103] George, J., Blanford, S., Domingue, M. J., Thomas, M. B., Read, A. F., Baker, T. C. Reduction in host-finding behaviour in fungusinfected mosquitoes is correlated with reduction in olfactory receptor neuron responsiveness. *Malar. J.* 2011, *10*, p. 1- 13, doi.org/10.1186/1475-2875-10-219.

[104] Lai, Y., Chen, H., Wei, G., Wang, G., Li, F., Wang, S. In vivo gene expression profiling of the entomopathogenic fungus *Beauveria bassiana* elucidates its infection stratagems in *Anopheles* mosquito. *Sci. China Life* Sci. 2017, *60*, p. 839-851, doi: 10.1007/s11427-017-9101-3.

[105] Rhodes, V. L, Thomas, M. B., Michel, K. The interplay between dose and immune system activation determines fungal infection outcome in the African malaria mosquito, *Anopheles gambiae. Dev. Comp. Immunol*. 2018, *85*, p. 125-133, doi.org/10.1016/j.dci.2018.04.008.

[106] Ishii, M., Kanuka, H., Badolo, A., Sagnon, N'F., Guelbeogo, W. N., Koike, M., Aiuchi, D. Proboscis infection route of *Beauveria bassiana* triggers early death of *Anopheles* mosquito. *Sci. Rep.* 2017, *7*, p. 1-10, doi:10.1038/s41598-017-03720-x.

[107] Howard, A. F. V., Koenraadt, C. J. M., Farenhorst, M. Knols, B. G. J., Takken, W. RPeyserarechthroid resistance in *Anopheles gambiae* leads to increased susceptibility to the entomopathogenic fungi *Metarhizium anisopliae* and *Beauveria bassiana*. *Malar. J.* 2010, *9*, p. 2-9, doi.org/10.1186/1475-2875-9-168.

[108] Howard, A. F. V., Guessan, R. N., Koenraadt, C. J. M., Asidi, A., Farenhorst, M., Akogbéto, M., Knols, B. G. J., Takken, W. First report of the infection of insecticide-resistant malaria vector mosquitoes with na entomopathogenic fungus under field conditions. *Malar. J.* 2011, *10*, p. 2-8, doi.org/10.1186/1475-2875-10-24.

[109] Hancock, P. A. Combinando biopesticidas fúngicos e mosquiteiros tratados com inseticida para melhorar o controle da malária. *Plos. Comput. Biol.* 2009, *5*, e1000525.

[110] Farenhorst, M., Knols, B. G. J., Thomas, M. B., Howard, A. F. V., Takken, W., Rowland, M., N'Guessan, R. Synergy in Efficacy of Fungal Entomopathogens and Permethrin against West African Insecticide-Resistant *Anopheles gambiae* Mosquitoes. *PLoS ONE* 2010, *5*, doi:10.1371/journal.pone.0012081.

[111] Singh, G., Prakash, S. Virulency of *Verticillium* sp. against mosquito vectors for malaria, filarial, and dengue. *Asian Pac. J. Trop. Dis.* 2015, *5*, S27-S30, doi:10.1016/S2222-1808(15)60850-7.

[112] Soni, N., Prakash, S. Larvicidal effect of *Verticillium lecanii* metabolites on Culex quinquefasciatus and *Aedes aegypti* larvae. *Asian Pac. J. Trop. Dis*. 2012, p. 220-224.

[113] Rodrigues, J., Camposa, V. C., Humber, R. A., Luz, C. Efficacy of Culicinomyces spp. against *Aedes aegypti* eggs, larvae and adults. *J. Invertebr. Pathol*. 2018, *157*, p. 104-11, doi.org/10.1016/j.jip.

[114] Singh, G., Prakash, P. Evaluation of culture filtrates of *Culicinomyces clavisporus*: Mycoadulticide for *Culex quinquefasciatus*, *Aedes aegypti* and *Anopheles stephensi. Parasitol. Res* 2012, 110, 267-272.

[115] Luz, C., Mnyone, L. L., Sangusangu, R., Lyimo, I. N., Rochaa, L. F. N., Humber, R. A., Russell, T. L. A new resting trap to sample fungus-infected mosquitoes, and the pathogenicity of *Lecanicillium muscarium* to culicid adults. *Acta Trop.* 2010, *116*, p. 105-107, doi:10.1016/j.actatropica.2010.05.001.

[116] Ramirez, J. L., Muturi, E. J., Dunlap, C., Rooney, A. P. Strain-specific pathogenicity and subversion of phenoloxidase activity in

the mosquito *Aedes aegypti* by members of the fungal entomopathogenic genus *Isaria*. *Sci. Rep.* 2018, *8*, p.1-12, doi:10.1038/s41598-018-28210-6.

[117] Pelizza, S. A., Lastra, C. C. L., Becnel, J. J., Humber, R. A., García, J.J. Further research on the production, longevity and infectivity of the zoospores of *Leptolegnia chapmanii* Seymour (Oomycota: Peronosporomycetes). *J. Invertebr. Pathol.* 2008, *98*, p. 314-319.

[118] Pelizza, S. A., Lastra, C. C. L., Maciá, A., Bisaro, V. Efecto de la calidad del agua de criaderos de mosquitos (Diptera: Culicidae) sobre la patogenicidad e infectividad de las zoosporas del hongo *Leptolegnia chapmanii* (Straminipila: Peronosporomycetes). *Rev. Biol. Trop.* 2009, *57*, p. 371-379.

[119] Pelizza, S. A., Scorsetti, A. C., Bisaro, V., Lastra, C. C. L., García, J.J. Individual and combined effects of Bacillus thuringiensis var. israelensis, temephos and *Leptolegnia chapmanii* on the larval mortality of *Aedes aegypti*. *BioControl* 2010, *55*, p. 647-656, doi: 10.1007/s10526-010-9281-2.

[120] Leles, R. N., Lastra, C. C. L., García, J. J., Fernandes, E. K. K., Luz, C. A simple method for the detection of *Leptolegnia chapmanii* from infected *Aedes aegypti* larvae. *Can. J. Microbiol.* 2013, *59*, p. 425-429, dx.doi.org/10.1139/cjm-2012-0703.

[121] Pelizza, S. A., Scorsetti, A. C., Tranchida, M. C. The sublethal effects of the entomopathic fungus *Leptolegnia chapmanii* on some biological parameters of the dengue vector *Aedes aegypti*. *J. Insect. Sci.* 2013, *13*, p. 1-8.

[122] Montalva, C., Santos, K., Collier, K., Rochaa, L. F. N., Fernandes, H. K. K., Castrillo, L. A., Luz, C., Humber, R. A. First report of *Leptolegnia chapmanii* (Peronosporomycetes: Saprolegniales) affecting mosquitoes in central Brazil. J. *Invertebr. Pathol.* 2016, doi: dx.doi.org/10.1016/j.jip.2016.03.012.

[123] Gutierrez, A. C., Rueda Páramo, M. E., Falvo, M. L., López Lastra, C. C., García, J. J., *Leptolegnia chapmanii* (Straminipila: Peronosporomycetes) as a future biorational tool for the control of

Aedes aegypti (L.). *Acta Trop.* 2017, doi:dx.doi.org/10.1016/j.actatropica.2017.01.021.

[124] Velasquez, P. F. Q., Abiff, S. K., Fins, K. C., Conway, Q. B., Salazar, N. C., Delgado, A. P., Dawes, J. K. Douma, L. G. Tartar, A. Transcriptome Analysis of the Entomopathogenic Oomycete *Lagenidium giganteum* Reveals Putative Virulence Factors. *Appl. Environ. Microbiol.* 2014, *80*, p. 6427-6436, doi:10.1128/AEM.02060-14.

[125] Maldonado-Blanco, M. G., Leal-López, E. Y., Ochoa-Salazar, O. A., Elías-Santos, M. Galán-Wong, L. J., Quiroz-Martínez, H. Effects of culture medium and formulation on the larvicidal activity of the mosquito pathogen *Lagenidium giganteum* (Oomycetes: Lagenidiales) against *Aedes aegypti*. *Acta Trop.* 2011, *117*, p. 114-118, doi.org/10.1016/j.actatropica.2010.10.010.

[126] Vilela, R., Taylor, J. W., Walker, E. D., Mendoza, L. *Lagenidium giganteum* Pathogenicity in Mammals. *Emerg Infect Dis*. 2015, *21*, p. 290-297, doi.org/10.3201/eid2102.141091.

[127] Singh, G., Prakash, S. Efficacy of *Lagenidium giganteum* (Couch) metabolites for control *Anopheles stephensi* (Liston) a malaria vector. *Malar. J.* 2010, *9*, p. 1-2.

[128] Rick, E. M., Woolnough, K., Pashley, C. H., Wardlaw. A. J. Allergic Fungal Airway Disease. *J. Investig. Allergol. Clin. Immunol.* 2016, 26, p. 344-354, doi: 10.18176/jiaci.0122.

[129] Twaroch, T. E., Curin, M., Valenta, R., Swoboda, I. Mold Allergens in Respiratory Allergy: From Structure to Therapy. *Pediatr. Allergy. Respir. Dis*. 2015, *7*, p. 205-220, dx.doi.org/10.4168/aair.2015.7.3.205.

[130] Ravindran, K., Akutse, K. S., Sivaramakrishnan, S., Wang, L. Determination and characterization of destruxin production in *Metarhizium anisopliae* Tk6 and formulations for *Aedes aegypti* mosquitoes control at the field level. *Toxicon* 2016, *120*, p. 89-96, dx.doi.org/10.1016/j.toxicon.2016.07.016.

[131] Maldonado-Blanco, M. G., Gallegos-Sandoval, J. L. Fernández-Peña, G., Sandoval-Coronado, C. F., Elías-Santos, M. Effect of

culture medium on the production and virulence of submerged spores of *Metarhizium anisopliae* and *Beauveria bassiana* against larvae and adults of *Aedes aegypti* (Diptera: Culicidae), *Biocontrol Sci. Technol.* 2014, *24*, p. 180-189, doi: 10.1080/09583157.2013. 855164.

[132] Scholte, E. J., Takken, W., Knols, B. G. J. Infection of Adult *Aedes aegypti* and *Ae. albopictus* Mosquitoes with the Entomopathogenic Fungus *Metarhizium anisopliae*. *Acta Trop.* 2007, *102*, p. 151-158, doi.org/10.1016/j.actatropica.2007.04.011.

[133] Garrido-Jurado, I., Alkhaibari, A., Williams, S. R., Oatley-Radcliffe, D. L., Quesada-Moraga, E., Butt, T. M. Toxicity testing of Metarhizium conidia and toxins against aquatic invertebrates. *J. Pest. Sci*. 2016, *89*, 557-564, doi: 10.1007/s10340-015-0700-0.

[134] Khodyrev, V. P., Dubovskiy, I. M., Kryukov, V. Yu., Glupov, V. V. Susceptibility of *Anopheles messeae* Fall. and *Culex pipiens pipiens* L. Larvae to Entomopathogenic Fungi *Metarhizium. Contemp. Probl. Ecol.* 2014, *7*, p. 334-337.

[135] Belevich, O., Yurchenko, Y., Krivopalov, A., Kryukov, V., Glupov, V. Effects of *Metarhizium robertsii* on the bloodsucking mosquito *Aedes flavescens* and non-target predatory insects (Odonata). *J. Appl. Entomol.* 2018, *142*, p. 632-635, doi: 10.1111/jen.12509.

[136] Valero-Jiménez, C. A., Debets, A. J. M., Kan, J. A. L. V., Schoustra, S. E., Takken, W., Waan, B. J. Z., Koenraadt, C. J. M. Natural variation in virulence of the entomopathogenic fungus *Beauveria bassiana* against malaria mosquitoes. *Malar. J.* 2014, *13*, p. 1-8, doi.org/10.1186/1475-2875-13-479.

[137] Sternberg, E. D., Ng'habi, K. R., Lyimo, I. N., Kessy, S. T., Farenhorst, M., Thomas, M. B., Knols, B. G. J., Mnyone, L. L. Eave tubes for malaria control in Africa: initial development and semi-field evaluations in Tanzania. *Malar. J.* 2016, *15*, p. 2-11, doi:10.1186/s12936-016-1499-8.

[138] Seye, F., Faye, O., Ndiaye, M., Njie, E., Afoutou, J. M. Pathogenicity of the fungus, *Aspergillus clavatus*, isolated from the locust, Oedaleus senegalensis, against larvae of the mosquitoes

Aedes aegypti, *Anopheles gambiae* and *Culex quinquefasciatus. J. Insect Sci.* 2009, *9*, p. 2-7.

[139] Seye, F., Bawin, T., Boukraa, S., Zimmer, J. I., Ndiaye, M., Delvigne, F., Francis, F. Pathogenicity of *Aspergillus clavatus* produced in a fungal biofilm bioreactor toward *Culex quinquefasciatus* (Diptera: Culicidae). *J. Pestic. Sci.* 2014, *39,* p. 127-132, 10.1584/jpestics.D14-006.

[140] Bawin, T., Seye, F., Boukraa, S., Zimmer, J-Y., Raharimalala, R., Zune, Q., Ndiaye, M., Delvigne, F., Francis, F. Production of two entomopathogenic *Aspergillus* species and insecticidal activity against the mosquito *Culex quinquefasciatus* compared to *Metarhizium anisopliae*. *Biocontrol Sci. Techn.* 2016, doi: 10.1080/09583157.2015.1134767.

[141] Ragavendran, C., Srinivasan, R., Kim, M., Natarajan, D. *Aspergillus terreus* (Trichocomaceae): A Natural, Eco-Friendly Mycoinsecticide for Control of Malaria, Filariasis, Dengue Vectors and Its Toxicity Assessment Against an Aquatic Model Organism *Artemia nauplii*. *Front. Pharmacol.* 2018, *9*, p. 1-18, doi: 10.3389/fphar.2018.01355.

[142] Ragavendran, C., Natarajan, D. Insecticidal potency of Aspergillus terreus against larvae and pupae of three mosquito species *Anopheles stephensi*, *Culex quinquefasciatus*, and *Aedes aegypti*. *Environ. Sci. Pollut. Res*. 2015, *22*, 17224-17237, doi:10.1007/s11356-015-4961-1.

[143] Singh, G., Prakash, S. Lethal Effects of *Aspergillus niger* against Mosquitoes Vector of Filaria, Malaria, and Dengue: A Liquid Mycoadulticide. *Sci. World J.* 2012, *2012*, p. 1-5, doi:10.1100/2012/603984.

[144] Vivekanandhana, P., Karthia, S., Shivakumara, M. S., Benellib, G. Synergistic effect of entomopathogenic fungus *Fusarium oxysporum* extract in combination with temephos against three major mosquito vectors. *Pathog. Glob. Health* 2018, *112*, 37-46, doi.org/10.1080/20477724.2018.1438228.

[145] Vivekanandhana, P., Deep, S., Kweka, E. J., Shivakumar, M. S. Toxicity of Fusarium oxysporum-VKFO-01 Derived Silver

Nanoparticles as Potential Inseciticide Against Three Mosquito Vector Species (Diptera: Culicidae). *J. Clust. Sci.* 2018, *29*, p. 1139-1149, doi.org/10.1007/s10876-018-1423-1.

[146] Mohanty, S. S., Raghavendra, K., Rai, U., Dash, A. P. Efficacy of female *Culex quinquefasciatus* with entomopathogenic fungus *Fusarium pallidoroseum*. *Parasitol. Res.* 2008, *103*, 171-174, 10.1007/s00436-008-0946-z.

[147] Banu, A. N., Balasubramanian, C. Optimization and synthesis of silver nanoparticles using *Isaria fumosorosea* against human vector mosquitoes. *Parasitol. Res.* 2014, *113*, p. 3843-3851, doi:10.1007/s00436-014-4052-0.

INDEX

A

adults, 4, 6, 40, 41, 48, 50, 53, 88, 92, 102, 104, 112, 114, 117
Aedes aegypti, v, vii, viii, 1, 2, 3, 4, 5, 6, 7, 8, 9, 10, 13, 14, 15, 16, 17, 18, 19, 20, 21, 22, 23, 24, 25, 33, 34, 37, 38, 39, 41, 43, 44, 46, 56, 57, 62, 63, 64, 65, 67, 68, 69, 70, 72, 75, 78, 101, 102, 103, 104, 105, 106, 107, 108, 112, 113, 114, 115, 116, 117, 118
Africa, 3, 15, 34, 37, 38, 46, 48, 50, 51, 52, 66, 67, 71, 73, 82, 87, 89, 117
agencies, 8, 10, 11
allergic reaction, 98
antibiotic, 57
aquatic habitats, 35, 40, 87
arboviral diseases, 34, 37, 44, 53, 62
Argentina, 19, 101
arthralgia, 16
arthritis, 22
arthropods, 89
Asia, 3, 23, 38, 47, 50, 107
Asian countries, 62
Aspergillus terreus, 118

B

bacteria, 57
bacterium, 14, 57
basophils, 98
benefits, 3, 55
biodegradability, 78
biodiversity, 54
bioindicators, 26
biological control, vii, ix, 3, 7, 50, 56, 78, 79, 80, 81, 82, 83, 88, 91, 96, 97, 108
biotic, 5
biotic factor, 5
birds, 54, 57
blood, 3, 6, 16, 17, 19, 28, 35, 39, 41, 42, 43, 45, 49, 79, 100
blood supply, 79
bone, 49
Brazil, 10, 19, 22, 25, 51, 58, 64, 65, 66, 70, 77, 82, 88, 99, 115
breathing, 36, 59
breeding, viii, 7, 10, 33, 34, 40, 41, 42, 50, 53, 57, 58, 59, 60, 66, 71, 86, 110
Burkina Faso, 46, 49, 51, 107

C

carbon, 59, 87
carbon dioxide, 59, 87
challenges, 65, 68, 72, 80, 87
chemical, viii, ix, 2, 6, 7, 9, 12, 34, 37, 50, 53, 57, 58, 78, 82, 87, 90, 95, 98, 99
chemicals, 13, 56
climate, 4, 20, 38, 67
climate change, 4, 67
climates, vii, 1, 2, 39
clinical trials, viii, 2, 7, 30
Colombia, 21, 51
colonization, vii, viii, 3, 33
combined effect, 89, 115
commercial, 58, 98
communication, 7, 12
communication strategies, 7
community, 7, 10, 12, 18, 22, 53, 56
compounds, 37, 55, 57, 72
containers, vii, 2, 5, 7, 21, 34, 36, 39, 41, 42, 53, 62
contamination, 112
control measures, viii, 2, 12
cost, 14, 18, 54, 58, 90
Côte d'Ivoire, 49, 51
Cuba, 49, 63
culture, 86, 114, 116, 117
culture media, 86
culture medium, 116, 117
cuticle, 40, 79, 85, 88, 90, 95, 105, 107, 112

D

damages, iv, 89
death rate, 97
deaths, 15, 46, 48, 62
dengue, vii, 1, 2, 3, 8, 9, 10, 13, 14, 15, 16, 17, 18, 19, 20, 21, 22, 23, 24, 25, 36, 38, 44, 47, 48, 49, 52, 58, 61, 62, 63, 64, 65, 66, 68, 69, 70, 71, 72, 73, 74, 78, 82, 90, 95, 102, 103, 104, 107, 108, 114, 115
dengue fever, viii, 2, 3, 8, 36, 48, 49, 58, 69
dengue hemorrhagic fever, 48, 49, 66
desiccation, 5, 39, 62, 94
destruction, 6, 97
detection, 28, 115
detoxification, 103
developed countries, 53, 54
developing countries, 9, 11, 12, 54
diseases, 2, 13, 25, 29, 34, 36, 37, 44, 48, 53, 56, 62, 64, 65, 67, 74, 78, 79, 80
dispersion, vii, 1, 2, 3, 22, 43, 98
distribution, 3, 10, 21, 42, 44, 51, 61, 65, 79
domestication, 23, 70
Dominican Republic, 24
dosage, 93, 94

E

ecosystem, 57
egg, 5, 16, 35, 36, 39, 41, 42, 45, 54, 59, 62, 64, 88
Egypt, 101, 111
encephalitis, 20, 30, 44
entomopathogenic fungi, v, vii, ix, 65, 77, 78, 79, 80, 81, 86, 87, 88, 93, 95, 97, 101, 104, 105, 107, 108, 109, 111, 112, 113, 117
environment, vii, 2, 6, 7, 8, 9, 10, 16, 37, 42, 82
environmental conditions, 36, 81, 82, 89
environmental contamination, 57
environmental effects, 55, 87
environmental factors, 4
environmental impact, 59
environmental management, 7, 8, 53
Environmental Protection Agency, 68, 90, 97
epidemic, 7, 10, 13, 19, 22, 61, 65, 67
epidemiology, 26, 66

ethical implications, 7
Europe, 3, 38, 45, 48, 52, 63, 65
evidence, 13, 18, 29, 47, 49, 61, 63, 85
evolution, 64, 81, 102
exposure, 79, 86, 87, 88, 93, 94, 98, 102, 109
extracts, ix, 34, 56, 69

F

fever, ix, 15, 22, 34, 35, 36, 44, 45, 46, 47, 48, 49, 52, 62, 63, 66, 68, 72, 73, 74, 110
field trials, 14
films, 28, 53
fish, 50, 53, 56, 60
France, 52, 66, 69
fungal infection, 79, 88, 90, 92, 93, 95, 112, 113
fungi, vii, ix, 50, 53, 57, 65, 78, 79, 80, 81, 82, 85, 86, 87, 88, 90, 93, 94, 95, 97, 98, 101, 104, 105, 106, 107, 108, 109, 112, 113
fungus, 79, 80, 82, 85, 86, 87, 88, 89, 90, 92, 93, 94, 95, 100, 102, 103, 104, 105, 106, 107, 108, 109, 111, 112, 113, 114, 115, 117, 118, 119

G

garbage, 12
gastrointestinal tract, 98
gene expression, 113
genes, 18, 90, 93, 97, 98, 112
genetic engineering, 98
genus, 34, 38, 45, 50, 51, 79, 81, 82, 83, 87, 88, 89, 91, 94, 95, 115
germination, 79, 90, 93, 104
global trade, 44
governments, 7, 11, 18
growth, vii, 2, 4, 9, 17, 42, 48, 53, 88, 90, 93, 111
growth hormone, 53
growth rate, 93

H

habitat, viii, ix, 2, 6, 26, 34, 36, 37, 39, 59, 60, 67
health, viii, 7, 10, 34, 48, 50, 53, 54, 69, 71, 86, 94
health education, 7
health effects, 94
health problems, 87
hemisphere, vii, 1, 2
history, 23, 38, 45, 66
homes, ix, 10, 12, 13, 34, 98
horizontal transmission, 45
host, 29, 45, 80, 85, 88, 90, 95, 104, 113
housing, 4, 9
human, viii, 2, 3, 6, 13, 16, 17, 24, 25, 27, 29, 33, 34, 38, 40, 43, 44, 45, 46, 47, 48, 50, 53, 54, 56, 67, 72, 78, 82, 87, 94, 98, 119
human animal, 3
human health, 53, 78, 82, 98
human resources, 48
humidity, 4, 36, 86, 108
hygiene, 3, 72

I

identification, 61, 105
immersion, 36, 86
immune response, 90, 98
immune system, 98, 112, 113
impregnation, 86, 88
imprisonment, 26
in vitro, 104, 107
incidence, viii, 2, 9, 15, 47, 62, 86
incompatibility, 58
incubation period, 93
incubation time, 86

individuals, 9, 94, 98
infection, vii, 2, 9, 17, 25, 27, 28, 29, 30, 45, 46, 47, 48, 50, 52, 62, 64, 67, 72, 79, 80, 90, 98, 101, 103, 104, 105, 109, 110, 113, 114
insecticide, viii, 2, 10, 11, 13, 21, 37, 50, 56, 74, 79, 85, 99, 100, 103, 104, 110, 114
insects, viii, 2, 13, 14, 26, 40, 55, 57, 78, 79, 81, 89, 107, 117
international trade, 38
intervention, 8, 14
invertebrates, 40, 117

K

kill, 57, 58, 59, 82, 88, 93, 94, 95

L

laboratory tests, 87
larvae, 4, 5, 6, 7, 36, 39, 40, 41, 42, 53, 56, 57, 58, 59, 60, 68, 69, 72, 79, 82, 83, 85, 86, 87, 92, 94, 95, 97, 99, 100, 102, 103, 104, 105, 108, 111, 112, 113, 114, 115, 117, 118
larval stages, 40, 53
Latin America, 12, 19, 21
legs, 6, 35, 40, 88, 93
life cycle, viii, 4, 19, 22, 33, 41, 89
liver, 46, 49
liver damage, 46
livestock, ix, 34
local government, 13
longevity, 79, 97, 115

M

malaria, 3, 18, 28, 30, 31, 74, 78, 86, 87, 93, 94, 95, 101, 102, 105, 106, 107, 109, 110, 113, 114, 116, 117
mammals, viii, 6, 16, 34, 43
management, 6, 7, 10, 18, 26, 37, 42, 60, 61, 65, 74, 75, 82, 89
marsh, 26, 60, 74
mass, 39, 70, 78, 97, 111
Mediterranean, 3, 21, 48
meta-analysis, 13, 19
metabolites, 85, 114, 116
metamorphosis, 39
methodology, 112
Mexico, 18, 23, 49, 58, 82
microbial communities, 28
microorganisms, 5
microscopy, 28
migration, 14, 41, 48, 49
morbidity, vii, 2, 16, 37, 44, 62
mortality, vii, 2, 37, 44, 48, 79, 80, 85, 86, 87, 92, 109, 111, 115
mortality rate, 80, 85, 92
mosquito, vii, viii, 1, 2, 3, 4, 5, 6, 12, 13, 14, 15, 16, 17, 19, 20, 21, 23, 24, 26, 29, 33, 34, 35, 36, 37, 38, 39, 41, 42, 43, 44, 45, 49, 50, 51, 52, 53, 54, 55, 56, 57, 58, 59, 60, 61, 62, 64, 65, 66, 67, 68, 70, 71, 72, 73, 74, 78, 79, 81, 82, 88, 90, 93, 94, 95, 97, 98, 99, 100, 101, 104, 105, 106, 107, 108, 109, 110, 111, 112, 113, 114, 115, 116, 117, 118, 119
mosquito bites, vii, viii, 2, 33
mosquitoes, vii, viii, 2, 4, 5, 6, 7, 9, 11, 13, 14, 16, 17, 18, 21, 22, 25, 30, 33, 34, 35, 36, 37, 38, 40, 41, 42, 43, 44, 45, 46, 48, 49, 51, 53, 54, 55, 56, 57, 58, 59, 60, 61, 64, 65, 70, 71, 73, 80, 82, 86, 87, 88, 90, 93, 94, 97, 98, 100, 102, 103, 104, 105, 106, 107, 110, 113, 114, 115, 116, 117, 119
musculoskeletal, 22
mutations, 20, 55

N

nanoparticles, 102, 119
natural enemies, 56
natural habitats, 7
Netherlands, 38, 63
neuron response, 93
New Zealand, 67, 89
next generation, 49
Nigeria, 33, 35, 47, 49, 50, 51, 55, 65, 67, 69, 75
Nile, 20, 29, 30, 89
North America, vii, 2, 52
nuisance, vii, viii, 33

O

oil, ix, 34, 42, 59, 68, 69, 85, 86, 87, 94, 108
oxygen, 6, 36, 39, 40

P

parasites, 56, 106
pathogenesis, 79, 85
pathogens, ix, 24, 34, 56, 57, 79, 97, 107
Peru, 1, 9, 10, 11, 21, 22, 23, 24, 25, 26, 27, 28, 30
pharmaceuticals, 55
population, 11, 14, 17, 20, 31, 45, 46, 47, 48, 50, 51, 53, 55, 59, 82, 87
population density, 17
predators, 50, 56, 57, 60
preparation, iv, 10
prevention, 6, 10, 12, 13, 62, 72, 73, 78
protease inhibitors, 85
public health, vii, viii, ix, 2, 11, 17, 20, 29, 34, 44, 46, 52, 66, 67, 78, 79, 80, 81, 98, 99
public sector, 12

R

rainfall, 4
rainforest, 45
real estate, ix, 34
receptor, 93, 113
recognition, 90
recreational areas, ix, 34
regions of the world, 38, 48, 56
resistance, viii, 2, 5, 7, 10, 11, 12, 13, 14, 18, 20, 21, 22, 55, 78, 79, 99, 110, 113
resources, 9, 68, 74, 80
response, 6, 69, 74, 85, 93, 98, 106
response capacity, 93
risk, vii, 2, 9, 11, 17, 38, 46, 47, 48, 52, 54, 62, 63, 64, 65, 71
risk factors, 48, 52, 63
risk perception, 65

S

safety, 79, 97, 99, 101, 103
Saudi Arabia, 12, 20
showing, 50, 87, 88
silver, 35, 102, 119
solid waste, vii, 2, 4, 17, 53
South America, 10, 16, 46, 50, 73
Southeast Asia, 3, 12, 18, 48, 50, 52
species, vii, viii, ix, 16, 17, 28, 33, 34, 35, 36, 37, 38, 40, 41, 42, 43, 45, 51, 54, 55, 57, 59, 61, 62, 66, 67, 78, 79, 81, 82, 85, 86, 87, 88, 89, 90, 93, 94, 95, 96, 97, 98, 99, 109, 118
spore, 79, 87, 94, 98
storage, 36, 42, 53
surveillance, 9, 10, 11, 29, 30, 49, 51, 64, 65, 67, 69
survival, 24, 85, 87, 88, 90, 95, 106, 109
susceptibility, 17, 55, 69, 95, 110, 113
syndrome, 16, 48, 63
synthesis, 89, 102, 119

T

Tanzania, 88, 110, 117
target, 12, 13, 55, 56, 57, 67, 78, 81, 82, 88, 94, 98, 107, 117
techniques, 3, 61, 89, 93
temperature, 4, 5, 21, 22, 24, 36, 39, 41, 63, 80, 90, 93, 95
temperature dependence, 93
thrombocytopenia, 72
transmission, vii, 2, 9, 13, 15, 16, 17, 19, 21, 22, 24, 31, 35, 37, 38, 41, 43, 45, 46, 47, 48, 49, 50, 51, 53, 56, 62, 63, 64, 65, 71, 74, 78, 93

U

United Kingdom, 66
United States, vii, viii, 33, 39, 49, 52, 60, 64, 90
urban, vii, 2, 3, 9, 16, 17, 34, 36, 42, 43, 46, 47, 48, 82
urban areas, 9, 42, 43, 46

V

vector, vii, 1, 2, 3, 7, 8, 9, 10, 11, 12, 13, 15, 16, 17, 18, 19, 20, 21, 22, 23, 24, 25, 26, 27, 28, 30, 34, 36, 37, 44, 45, 47, 48, 49, 51, 53, 54, 57, 61, 62, 64, 65, 67, 68, 72, 73, 74, 75, 79, 80, 82, 85, 86, 88, 89, 90, 92, 94, 95, 97, 98, 101, 102, 103, 104, 105, 106, 107, 109, 110, 111, 114, 115, 116, 118, 119
vector control, viii, 1, 2, 7, 9, 10, 11, 12, 15, 19, 26, 27, 28, 34, 37, 51, 53, 54, 61, 62, 67, 72, 73, 74, 79, 90, 94, 98, 101
vertical transmission, 45
viral diseases, 45, 68, 70, 82
viral infection, 15
virology, 65
virus infection, 16, 65, 74
virus replication, 90
viruses, ix, 14, 17, 20, 34, 36, 38, 41, 44, 48, 57

W

water, vii, viii, 2, 3, 5, 6, 7, 12, 21, 33, 34, 36, 39, 40, 41, 42, 53, 58, 59, 60, 62, 74, 86, 88, 95
water supplies, 42
water vapor, 59
West Africa, 43, 66, 89, 100, 112, 114
West Nile fever, 16
Western countries, 52
World Health Organisation, 72, 73, 74, 75
worldwide, viii, 33, 34, 37, 44, 46, 72, 79

Y

yellow fever, 15, 17, 24, 36, 38, 46, 47, 49, 52, 62, 64, 66, 73, 78, 82